Hybrid Intelligence for Social Networks

Hema Banati • Siddhartha Bhattacharyya •
Ashish Mani • Mario Köppen
Editors

Hybrid Intelligence for Social Networks

Springer

Editors
Hema Banati
Dept. of Computer Science
Dyal Singh College
University of Delhi
Delhi, India

Ashish Mani
Dept. of Electrical & Electronics
  Engineering
Amity University
Noida
Uttar Pradesh, India

Siddhartha Bhattacharyya
Dept. of Information Technology
RCC Institute of Information Technology
Kolkata
West Bengal, India

Mario Köppen
Network Design and Research Center
Kyushu Institute of Technology
Fukuoka, Japan

ISBN 978-3-319-87955-0   ISBN 978-3-319-65139-2   (eBook)
DOI 10.1007/978-3-319-65139-2

Printed on acid-free paper

This Springer imprint is published by Springer Nature
The registered company is Springer International Publishing AG
The registered company address is: Gewerbestrasse 11, 6330 Cham, Switzerland

Dedication

Dr. Hema Banati would like to dedicate the book to her first social network on this earth: her precious family!

Prof. (Dr.) Siddhartha Bhattacharyya would like to dedicate this book to his late father, Ajit Kumar Bhattacharyya, his late mother Hashi Bhattacharyya, his beloved wife Rashni and his cousin brothers Subrata, Bishwarup, Somenath and Debasish

Dr. Ashish Mani would like to dedicate this book to Most Revered Dr. Prem Saran Satsangi, Chairman Advisory Committee on Education Dayalbagh

Prof. (Dr.) Mario Koeppen would like to dedicate this book to Prof. (Dr.) Albert-László Barabási, a pioneer in network theory research.

Preface

The field of social networks has assumed paramount importance because of its omnipresence in our daily lives. As social networks have become the backbone of our society, this book is aimed at bringing under a common umbrella the different dimensions/perceptions of social networks varying from applications to the development of new artificial intelligent techniques for social networks to understanding the impact of social networks in e-commerce.

The traditional way of handling these networks was focussed on three primary areas

- Conceptual: Mathematical models, novel algorithms, frameworks, etc.
- Analytical: Trend analysis, data analysis, opinion analysis, data mining, knowledge extraction, etc.
- Technological: Tools, innovative systems, applications, etc.

Recent focus has shifted to the application of artificial intelligence techniques for understanding social networks on the Web. These help in the handling of the shortcomings and limitations associated with the classical platforms of computation, particularly for handling uncertainty and imprecision. Soft computing as an alternative and extended computation paradigm has been making its presence felt. Accordingly, a phenomenal growth of research initiatives in this field is evident. Soft computing techniques include the elements of *fuzzy mathematics*, primarily used for handling various real-life problems engrossed with uncertainty; the ingredients of *artificial neural networks*, usually applied to cognition, learning and subsequent recognition by machines, thereby inducing the flavour of intelligence in a machine through the process of its learning; and components of *evolutionary computation* mainly used for searching, exploration and the efficient exploitation of contextual information and knowledge, which are useful for optimization.

Individually, these techniques have got their strong points in addition to limitations. In several real-life contexts, it has been observed that they play a supplementary role for one another. Naturally, this has given rise to serious research initiatives for exploring avenues of hybridization of the above-mentioned soft computing techniques. This resulted in more robust and intelligent solutions in the form of *neuro-fuzzy, fuzzy-genetic, rough-neuro, rough-fuzzy, neuro-fuzzy-genetic, neuro-fuzzy-rough, quantum-neuro-fuzzy* architectures. Interestingly, the scope of such hybridization is gradually found to be all- encompassing.

The book provides a multidimensional approach to social networks. The first two chapters deal with the basic strategies for social networks, such as mining text from such networks or applying social network metrics using a hybrid approach, the next six focus on the prime research areas in social networks: community detection, influence maximization and opinion mining. The chapters "Community Detection Using Nature-Inspired Algorithms," "A Unified Framework for Community Structure Analysis in Dynamic Social Networks," "GSO-Based Heuristics for Identification of Communities and Their Leaders," and "A Holistic Approach to Influence Maximization" present varied intelligent approaches to detecting communities and maximizing influence through evolutionary and other hybrid techniques. The chapters "Opinion Dynamics Through Natural Phenomenon of Grain Growth and Population Migration" and "Opinion Mining from Social Travel Networks" focus on mining opinions from social networks. The last five chapters of the book, "Facilitating Brand Promotion Through Online Social Media: A Business Case Study," "Product Diffusion Pattern Analysis Model Based on Users' Reviews of E-commerce Applications," "Hierarchical Sentiment Analysis Model for Automatic Review Classification for E-commerce Users," "Trends and Pattern Analysis in Social Networks," and "Extensible Platform of Crowdsourcing on Social Networking Sites: An Analysis," concentrate on studying the impact and use of social networks in society, primarily in education, the commercial sector and the upcoming technology of crowd sourcing.

The book begins with a study of the hybrid techniques used to mine textual data from social networks and media data in the chapter **"Hybrid Intelligent Techniques in Text Mining and Analysis of Social Networks and Media Data"**. It considers the fact that social networks are considered to be a profuse source of viewpoints and an enormous amount of social media data is produced on a regular basis. This chapter presents a detailed methodology on how data mining, especially text mining, is applied to social networks in general. The relevance of this is essential because of the huge amount of data generated from the communication between users signed up for the various social media platforms on several topics such as books, movies, politics, products, etc. The users vary in terms of factors such as viewpoints, scenarios, geographical situations, and many other settings. If mined efficiently, there is the potential to provide a helpful outcome of an exegesis of the social quirks and traits. Furthermore, it introduces the traditional models used in mining the various hybrid methodologies that have evolved and provides a comparative analysis.

Basic parameters for social network analysis are social network metrics. There are numerous social network metrics. During the data analysis stage, the analyst combines different metrics in the search for interesting patterns. This process can be exhaustive with regard to numerous potential combinations and how we can combine different metrics. On the other hand, other non-network measures can be observed along with social network metrics. The chapter **"Social Network**

Metrics Integration into Fuzzy Expert Systems and Bayesian Networks for Better Data Science Solution Performance" proposes a methodology on fraud detection systems in the insurance industry, where a fuzzy expert system and the Bayesian network were the basis for the analytical platform, and social network metrics were used as part of the solution to improve performance. The solution that was developed shows the importance of integrated social network metrics as a contribution to better accuracy in fraud detection.

Community detection in social networks has become a dominating topic of current data research as it has a direct implication for many different areas of importance, whether social networks, citation networks, traffic networks, metabolic networks, protein-protein networks or web graphs, etc. Mining meaningful communities in a real-world network is a hard problem because of the dynamic nature of these networks. The existing algorithms for community detection depend chiefly on the network topologies and are not effective for such sparse graphs. Thus, there is a great need to optimize these algorithms. Evolutionary algorithms have emerged as a promising solution for the optimized approximation of hard problems. The chapter **"Community Detection Using Nature-Inspired Algorithms"** proposes to optimize the community detection method based on modularity and NMI using the latest grey wolf optimization algorithm. The results demonstrate the effectiveness of the algorithms, compared with other contemporary evolutionary algorithms.

One of the major tasks related to the structural social network analysis is the detection and analysis of community structures, which is highly challenging owing to consideration of various constraints when defining a community. For example, community structures to be detected may be disjoint, overlapping, or hierarchical, whereas community detection methods may vary depending on whether links are weighted or unweighted, directed or undirected, and whether the network is static or dynamic. The chapter **"A Unified Framework for Community Structure Analysis in Dynamic Social Networks"** presents a unified social network analysis framework that is mainly aimed at addressing the problem of community analysis, including overlapping community detection, community evolution tracking and hierarchical community structure identification in a unified manner. The chapter also presents some important application areas wherein the knowledge of community structures facilitates the generation of important analytical results and processes to help solve other social network analysis-related problems.

The chapter **"GSO-Based Heuristics for Identification of Communities and Their Leaders"** focuses on identifying leaders in communities detected in social networks. It presents a framework-based approach for extraction of communities using a very efficient nature-based DTL-GSO heuristic optimization algorithm based on the group search optimization algorithm. Using the proposed approach, designated communities and their leaders facilitate the task of harmonizing community-specific assignments in a more efficient manner, leading to an overall performance boost. The significant contribution of this chapter is a novel metric to identify community-specific central nodes (leaders), which are also highly trusted globally in the overall network.

Interesting applications of social networks include viral marketing, recommendation systems, poll analysis, etc. In these applications user influence plays an important role. The chapter **"A Holistic Approach to Influence Maximization"** discusses how effectively social networks can be used for information propagation in the context of viral marketing. Picking the right group of users, hoping they will cause a chain effect of marketing, forms the core concept of viral marketing applications. The strategy used to select the correct group of users is the influence maximization problem. The chapter proposes one of the viable solutions to influence maximization. The focus is to find those users in the social networks who would adopt and propagate information, thus resulting in an effective marketing strategy. The three main components that would help in the effective spread of information in the social networks are the network structure, the user's influence on others and the seeding algorithm. An amalgamation of these three aspects provides a holistic solution to influence maximization.

Opinion dynamics, yet another aspect of social networks, has witnessed a colossal interest in research and development activity aimed at the realization of intelligent systems, facilitating and predicting understanding. Many nature-inspired phenomena have been used for modelling and investigation of opinion formation. One of the prominent models based on the concept of ferromagnetism is the Ising model in statistical mechanics. This model represents magnetic dipole moments of atomic spins, which can exist in any one of two states, +1 or −1. In the chapter **"Opinion Dynamics Through Natural Phenomenon of Grain Growth and Population Migration"** NetLogo is used to simulate the Ising model and correlates the results with opinion dynamics in its purview. The novelty of the chapter lies in the fact that the grain growth phenomenon has been investigated to analyze opinion dynamics, providing a new unexplored dimension of studying opinion dynamics. The chapter presents a modelling of the natural phenomenon of population growth using rigorous mathematics and corroborated the results with opinion dynamics.

A close correlation between opinion dynamics and sentiment analysis forms the basis of the chapter **"Opinion Mining from Social Travel Networks"**. This deals with the extraction of emotions expressed in natural language into digital form. It is being used in every field to monitor people's emotions and sentiments. The chapter studies and analyzes the emotions expressed by people about their travel experiences.

Online social networks promote faster propagation of new ideas and thoughts. The percolation of social media into our daily life has influenced the way in which consumers interact and behave across the world. The traditional media marketing tools are facing increased competition from the more attractive alternatives provided by social media. With more than two thirds of the internet population connected through online networking sites such as Facebook, Twitter and Myspace, the potential offered by this medium is tremendous. The chapter **"Facilitating Brand Promotion Through Online Social Media: A Business Case Study"** analyses the case studies of three brand promotion activities through Facebook and models the mechanics of successful diffusion in comparison with traditional channels.

Another interesting aspect of online e-commerce systems are the users' reviews and ratings of products through their daily interaction. The chapter **"Product Diffusion Pattern Analysis Model Based on Users' Reviews of E-commerce Applications"** studies product diffusion pattern analysis as an impact of users' reviews on social e-commerce giant Amazon as a function of time. The proposed user review-based product diffusion pattern analysis (PDPA) model extracts reviews and its associated properties, such as review ratings, comments on reviews and the helpfulness of reviews using Amazon application programming interfaces and assigning weight to all reviews according to the associated properties mentioned. The model predicts the long-term dynamics of the product on popular interactive e-commerce sites by analyzing the users' offered reviews and ratings on these sites. The introduced users' review feature-based rise and fall product diffusion pattern analysis model adds a sense of quality assurance for products and services, which is a new dimension of organic marketing.

Assessing the quality of the product becomes very important. Product review classification is used to analyze the sentiment from reviews posted by the user to prepare the product report. The chapter **"Hierarchical Sentiment Analysis Model for Automatic Review Classification for E-commerce Users"** proposes a mechanism for opinion mining over text review data to generate product review reports based upon multiple features combined. This report shows positive and negative points about the specific product, which can play a significant role in the selection of the products on the online portals. The evolutionary history of the human brain shows advancement in its complexity and creativity during its evolutionary path from early primates to hominids and finally to *Homo sapiens*. When a problem arises, humans make use of their intelligence and various methods to find the solution. No doubt, they have come up with the best solutions, but many questions have been asked about how that problem is approached and how the solution is derived. The peculiar thing is that everyone has a different mechanism of thinking and comes up with different patterns of solutions. Can this pattern be mimicked by a machine where a problem can be solved by inputs from multiple individuals?

The chapter **"Trends and Pattern Analysis in Social Networks"** addresses this issue using the concepts of crowd sourcing and neural networks. Crowd sourcing deals with the pooling of ideas by people. The more people, the wider the perspective we get. The data given by them is processed and the field of neural networks plays a vital role in analyzing the data. These data contain various patterns and hidden solutions to many problems.

Using social networking in higher education is yet another critical domain of social networks owing to the complex nature of serving the population of the online generation. Utilizing social networking environments for learning and teaching at HEIs could be a financially effective and productive way of speaking and connecting with online higher education members. The chapter **"Extensible Platform of Crowdsourcing on Social Networking Sites: An Analysis"** contributes in terms of analysing the technological behaviour patterns (their active or relevant engagement) of the online users on the higher education social media networks of the higher education institutions.

The experimental findings of each of the chapters reveal that the hybrid soft computing paradigms yielded superior performance compared with their traditional counterparts. This substantiates the fact that proper hybridization of the soft computing tools and techniques always leads to more effective and robust solutions, as the constituent soft computing elements in the hybrid system always complement each other.

The editors have tried to bring together some notable contributions in the field of hybrid intelligent techniques for social networks. The editors feel that these contributions will open up research interests to evolve more robust, time-efficient and fail-safe hybrid intelligent systems. The editors believe that this book will serve graduate students and researchers in computer science, electronics communication engineering, electrical engineering and information technology as a reference book and as an advanced textbook for some parts of the curriculum. Last but not the least, the editors would like to take this opportunity to extend their heartfelt thanks to the staff of Springer and to Mr. Ronan Nugent, Senior Editor, Springer for his constructive support during the tenure of the book project.

Delhi, India   Hema Banati
Kolkata, India   Siddhartha Bhattacharyya
Noida, India   Ashish Mani
Iizuka, Japan   Mario Köppen
May 2017

Contents

Hybrid Intelligent Techniques in Text Mining and Analysis of Social Networks and Media Data

Neha Golani, Ishan Khandelwal, and B.K. Tripathy

Abstract Text data from social media and networks are ubiquitous and are emerging at a high rate. Tackling these bulky text data has become a challenging task and an important field of research. The mining of text data and examining it with the help of several clustering techniques, classification techniques, and soft computing methods has been studied in a comprehensive manner in the past. This chapter focuses mainly on the hybrid techniques that have been used to mine textual data from social networks and media data. Social networks are considered a profuse source of viewpoints and outlooks of the public on a worldwide scale. An enormous amount of social media data is produced on a regular basis, generated because of the communication between the users who have signed up for the various social media platforms on several topics such as books, movies, politics, products, etc. The users vary in terms of factors such as viewpoints, scenarios, geographical situations, and many other settings. If mined efficiently, the data have the potential to provide a helpful outcome of an exegesis of social quirks and traits. This chapter offers a detailed methodology on how data mining, especially text mining, is applied to social networks in general. Furthermore, it goes on to introduce the traditional models used in mining the various hybrid methodologies that have evolved and make a comparative analysis. We also aim to provide the future scope and research studies present in this field with all possible new innovations and their applications in the real world.

Keywords Ant colony optimisation • Hybrid techniques • Neural networks • Particle swarm optimisation • Social networks • Support vector machine • Text mining

1 Introduction

A vast amount of research has been done with respect to techniques in Text Mining, Analysis of Social Networks and Media Data. An application has been created that intends to take the information in text form linked to the data that is pertinent to

N. Golani (✉) • I. Khandelwal • B.K. Tripathy
VIT University, Vellore, India
e-mail: nehagolani00@gmail.com; ishankhandelwal23@gmail.com; tripathybk@vit.ac.in

identifying the geographical location of a person or device by means of digital information processed via the Internet. These data are taken as input and data mining is performed in the settings where the network of the social media data is complex. The idea is to extract the information, which works on social recommendation. It has addressed the challenge of the inability of the recommender systems to refine the search and posit the suggestions to the application users. It also puts forth the concept of Social Network based on Location linked with methodologies of text mining, and discloses issues that still require research for more efficacious and integrated outcomes.

The opinion mining techniques are applied to social media data such as Twitter data, which proves to be an effective means of portraying the organisations in terms of speed and efficiency of delivering the message along with the wide range of audience covered. Several features and techniques for training the opinion classifiers for Twitter datasets have been studied in the past few years, with different results. The problems of the previous conventional techniques are the precision of classification, the data being sparse, and sarcasm, as they incorrectly categorise the Tweets in the opposite category or neutral owing to the failure to understand the previously mentioned factors.

Social media have had an extremely significant impact on our lives and on our interactions with others. This makes it necessary for a successfully working organisation to have the capability to analyse the current occurrences using the information available, which involves reviews and experiences of customers of the company's services and products to predict the near future. This would assist the company to develop a better customer experience and ultimately build a better stature for the company. Several companies do not have enough knowledge on the data mining of social media efficaciously and they are not acquainted with the competitive intelligence of social media. Because of the abilities of text mining, it can be concluded that the implementation of text data to social media data can provide fascinating insights into human interactions and psychology.

Moreover, one of the endowments of the research relates to the study of potential linguistic tags as they enable the service-users to develop a vocabulary of their own and eventually examine the investigated or mapped areas and impart knowledge on these. They aspire to extend this technique with the assistance of similarity metrics to build personalised recommendations on the Web using collaborative filtration of data in Data Mining, culminating in a hybrid outlook.

Soft computing (SC) constitutes various paradigms of computing, which involve neural networks (NNs), genetic algorithms (GAs), fuzzy logic (FL), which are helpful in generating substantial and effective hybrid intelligent systems for text mining. An immense amount of research has been done in the field of hybrid methods of text mining. Kumar et al. [13] proposed a neuro-fuzzy method for mining of social media data with the help of the Cloud. Figure 1 [22] depicts the basic methodology for the analysis and mining of big text data.

The process of text mining of the social media data can be categorised into three steps. The first step is the text pre-processing step in which the textual data are extracted from the social media sites, collected and stored as the preparation of text

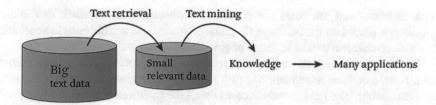

Fig. 1 Methodology for analysing big text data: text retrieval and text mining [22]

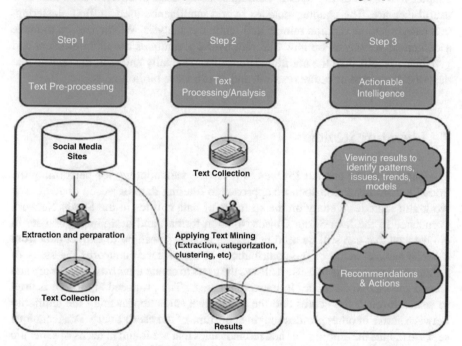

Fig. 2 Process of text mining for social media content [6]

processing and analysis. In the text processing/analysis step, an assortment of text mining techniques are applied to the data collected in the first step. The final step is the actionable intelligence step in which the results obtained from the previous steps are further processed to identify the various patterns, issues, trends, and models and to provide vital recommendations and actions based on the knowledge obtained (see Fig. 2).

The social network data are not very unusual from the conventional data, although a specialised language used for social media data to explain the skeleton and indices of the sets of observations that are usually taken into consideration. Furthermore, the datasets, which are developed by the networkers, generally result in different looking data array from the traditional rectangular data array with which the researchers and analysts are acquainted. One of the important differences between conventional data and network data is that with conventional data factors such as actors and attributes are considered, whereas with network data more emphasis is placed on the actors and relations. This creates a difference in approach that a researcher has to take in terms of generation of a research design.

The chapter has begun with a brief introduction on social network and media data. Next, a concise introduction is given on text mining, sentiment analysis, and opinion mining. The next section focuses on all the traditional methods being used and introduces an in-depth analysis of the mathematics required [1, 21]. Then, the chapter moves on to the hybrid techniques being used in social media and how useful they are. The chapter includes hybrid intelligence used in Text clustering, text categorisation, opinion mining and sentiment analysis. We then go on to make a comparative analysis on how different hybrid techniques are more efficient and give better results than the traditional concepts. We finally end with the section that describes the scope of future research and an extensive bibliography.

2 Literature Review

The purpose of the chapter [5] was to present a methodology for performing the Social Recommendation-focused approach to filtering data based on content. The work also includes a study on the structure of data entities in the Social Network (structured in the Java Script Object Notation format) and designing a crawler to collect data that can still be used in other approaches needing to extract data from a social network. One of the contributions of the research concerns the study of potential semantic tags because it allows the users to create their own vocabulary and spread knowledge discovery in unexplored areas. The proposed algorithm is used to search through the scores and the cumulative sums approaching the similarity between items, or other profiles sought at the time of the user's query. The technique seeks to reduce the ambiguities and redundancies that are found in terms of semantic relations. The main contribution of this work is undoubtedly the creation of a new methodology that could be adopted in the recommendation process on location based social networks web environment, where further research is still required in addition to consultation with the user and the use of data mining techniques in text, with consolidated results.

In [13], the authors have tried to analyse the advantages and pitfalls of artificial neural networks (ANN) and fuzzy approaches to mining social media datasets. They analysed Web mining and its types such as classification and clustering. It gives us an insight into Web usage mining through artificial neural networks, the use of fuzzy logic and ant colony optimisation (ACO) in web mining. It suggests the use of Social Network Analysis (SNA) as an essential tool for researchers owing to the increase in the number of social media users. It enlists merits and demerits of several methods in soft computing, such as genetic algorithm, artificial neural network, ant colony optimisation and fuzzy set for mining the datasets of social media.

In [10], the authors have implemented and evaluated Naïve Bayes and J48 algorithms under two-term frequency and TF-IDF (Term Frequency Inverse

Document Frequency) term weighting methods. They concluded that the J48 method performs better than the Naïve Bayes in terms of frequency and TF-IDF term weighting methods. It is concluded that in text classification, the J48 method outperforms Naïve Bayes in term frequency and TF-IDF term weighting methods. Also, according to results obtained, TF-IDF has better performance than the term frequency method in text classification.

Furthermore in [15], to categorise each opinion given by the viewers as positive or negative for a particular review dataset, hybrid methods are put forward for consideration. The reviews and ratings of movies on Twitter are considered in Twitter with the help of sentiment analysis. A hybrid method using particle swarm optimisation (PSO) and support vector machine (SVM) is used to categorise the opinions of the user into positive, negative, for a remarks dataset of a particular movie. These results are helpful in an improved decision-making process.

In [12], the authors have proposed a novel hybrid approach for determining the sentiment of individual Tweets on Twitter. They proposed a framework that uses a hybrid design to combine three kinds of classification algorithms for enhanced emoticon analysis, SentiWordNet analysis and polarity classifier. The proposed framework consists of three main modules for data acquisition, preprocessing of data and classification algorithms pipelined to classify the Tweets into positive, negative or neutral. The datasets were generated and the experiments were conducted on six different datasets consisting of random tweets acquired from Twitter using Twitter streaming API. Experimental conclusions mentioned in this chapter conclude that the hybrid methods perform significantly better than the individual components. Thus, the results show that these hybrid techniques end up with more accurate results compared with similar techniques individually applied to the datasets. They achieved an average efficiency of 85.7% with 85.3% precision and 82.2% recall while using the hybrid technique. They also significantly contributed to decreasing the number of Tweets categorised as neutral. The frameworks tested showed enhancement in precision, effectiveness and recall when hybrid techniques were used.

3   Social Network and Media

A social network is a complex system of individuals and or organisations. It is an online platform that enables people to forge social contacts and relations with other people and organisations that harbour similar interests. Today, a plethora of social-networking websites are available to internet users that provide an array of social media services such as e-mail, instant messaging, online forums, blogging etc. They often allow users to share their ideas and thoughts through comments, personalised messages, broadcast messages, public posts, digital photos, videos, and through audio. A variety of social networking websites are easily accessible to users through internet enabled desktops, laptops, tablet computers plus smartphones and smart-watches. They keep the users up to date with real-world activities and events happening within their social network.

Since their inception, the social networking websites have amassed millions of active users. The following are some of the popular and influential social networking websites and mobile applications:

- Facebook: A social networking website and service that allows the users to stay connected with their friends and family. Users can post comments, share photographs and videos and follow pages of interest.
- Twitter: A social networking service that allows users to read and share short messages (maximum 140 characters) called Tweets.
- YouTube: A video sharing and rating website that allows user to upload, view, rate, comment on and share videos. It is the world's largest video platform.
- WhatsApp: An instant-messaging service for smartphones that enables users to send text messages, images, audio and video in real-time.
- LinkedIn: A social networking service that enables a user to build professional networks and contacts.

4   Text Mining

Text mining [3] is the discovery and exploration of hidden information by processing the textual data from various sources. The objective of text mining is to find interesting patterns from large databases. Thus, it gleans useful information from text and processes it to generate important trends through pattern learning. Text mining structures the input text by parsing and adding or removing derived linguistic features, and inserting the subsequent features into the database.

Text is the preferred method of communication in the social media. It is extremely vital to extract the information and knowledge from the profusion of text data extracted from social media and networking websites. The processing of this natural language text extracted from social media has led to the discovery of complex lexical and linguistic patterns that can assist and aid in the answering of complex questions and form an important knowledge base [8]. Thus, text mining has huge potential in the commercial and in the research and educational sectors, as the unstructured and unprocessed text is the largest easily available source of knowledge. Refined data mining techniques are needed to extract the unknown tacit and potentially useful information from the data [9].

The process of text mining can be subdivided into two distinct phases. The first phase is the text refining phase in which the transformation of free text into an intermediate form takes place. Various data mining techniques can be directly applied on the intermediate form. The intermediate form can be represented in various forms such as conceptual graphs, relational data representation and object-oriented. The second phase is the knowledge distillation form. In this phase, data mining techniques such as clustering and classification are applied to the

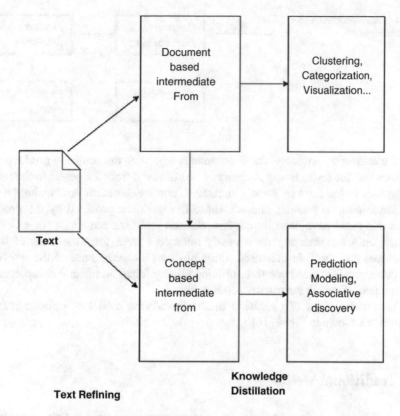

Fig. 3 General framework for text mining

intermediate form with the intention of inferring important patterns that could be incorporated into the knowledge database (refer Fig. 3) [19].

5 Opinion Mining and Sentiment Analysis

Opinion mining refers to the outlook or viewpoint of every user on certain products, services, topics, etc. Sentimental Analysis can be broadly applied to social media for the purposes of communicating, delivering and exchanging offerings that have value for the public in general and for managing the relationship between client and customer. For example, viewers post many observational comments on social media networks about news feeds, or they express their judgement and remarks on current affairs. These remarks made by each individual can be taken as opinions on the shared data that they post on these web forums [16].

The application of opinion mining would be for organisations, clients etc. to examine the reviews and evaluate the viewers and users towards their organisations

Fig. 4 Framework model for opinion mining

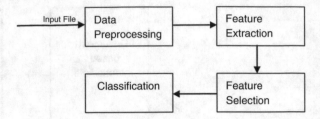

and the respective products. These comments regarding the company products can be given and the respective messages can be conveyed from the public to the social media sites in the form of Tweets, messages, comments etc. Several techniques of opinion mining to evaluate and scrutinise the feedback provided by the product or service users have been introduced. Several real-time scenarios, for example, reviews on a documentary for a newly released movie, form the basis of these techniques and result in increased profitability as a consequence of the research, in a commercial sense. Therefore, opinion mining forms an integral component of top companies, which are usually product-based.

The various steps of execution in the foundation model of opinion mining systems are shown in Fig. 4 [16].

6 Traditional Methods

The traditional techniques are the stand-alone techniques used for the text mining of social media data. They operate independently and cannot be sub-divided, unlike the hybrid techniques, which are made up of various composite methods. There are many traditional techniques that are used to mine the social media data. These techniques are broadly classified into machine learning-based and lexicon-based techniques.

The Machine learning (ML) techniques are mainly the supervised classification methods and therefore need two types of datasets: a training dataset and a test dataset. A supervised learning classifier uses the training dataset to learn the diverse and distinctive patterns present in the data, whereas the test dataset is used to examine the performance of the classifier. The lexicon-based techniques consist of unsupervised techniques in which the features of the texts are analysed with the lexicons having predetermined values. Some of the prominent machine learning techniques used for the mining and pattern extraction of social media data are discussed in the following sections.

6.1   Soft Computing for Web Mining

Soft computing is different from conventional (hard) computing, as, unlike hard computing, it is tolerant of inaccuracy, ambiguity, incomplete truth and approximation. It is an amalgamation of methods and techniques that operate collectively and have an impact in one form or the other on adaptable information processing for handling situations that are obscure and cryptic in real life. It is aimed at deriving benefit by utilising the endurance for inaccurate, incertitude, approximate reasoning and the partial truth to obtain tractability, resilience, relatively inexpensive solutions and close congruence to the decision-making mechanism, like that of human beings.

The principle that steers it is the formulation of the computational approach that would lead to a solution that is admissible at low cost by seeking an approximate solution to an inaccurately or accurately devised problem. There is a requirement for incorporating and implanting agent-based intelligent systems into the web tools to facilitate web intelligence. Designing of intelligent systems (both client and server-side) requires the scrutiny of the researchers from various domains such as artificial intelligence, ML, knowledge discovery etc. Such systems are capable of mining knowledge both in specific web zones and across the Internet.

Although the issue of designing automated tools to detect, extract, reduce, filter and access information that the users demand from unlabelled scattered, and assorted web data has been an unsolved mystery to date, soft computing seems like a viable option for managing these traits and properties, and for dealing with some of the constraints of existing techniques and technologies [15].

6.2   ANN for Web Usage Mining

In ML, an artificial neural network (ANN) is a nexus kindled by biological neural networks that estimates certain functions by considering a large number of inputs, usually unknown. The fundamental component of this information-processing structure is a large amount of interconnected processing components known as neurons working in unison to deal with a particular problem. An ANN is arranged in a certain configuration to work for a specific application, such as recognition of patterns or data classification, through a learning procedure.

There are various advantages of an ANN that make it well suited to web mining and analysis of social media data. An ANN is an adaptive learning technique that organises and coordinates the assignment of humans to produce the desired effect as per the requirements and demands of the users with the help of the computers. An ANN also has the ability to generate its own arrangement or depiction of the information it receives during the phase of learning. This property is termed self-organisation maps (SOMs). In addition, ANN computations can also be carried out in parallel and are fault-tolerant via Redundancy Information Coding [20].

The source of web log data taken into account for the assessment can include any specific web server for a particular time limit. Initially, data cleaning is carried out to get rid of the redundant and irrelevant log that has a potential to slow down the process of obtaining the scheme and trend of web usage. Once the data are refined, the user identification process is carried out with the help of an IP address and user's agent fields. By assisting with applying the algorithms, the users are uniquely and distinctly recognised and the path of sessions whose transactions have been accomplished are obtained.

The URL clustering is achieved with the help of K-means to obtain the frequency of each URL in each cluster. Clustering is a technique used to partition the data elements into clusters keeping similar data in the same cluster and finding unknown patterns in the datasets. Each cluster is denoted as an adaptively varying centroid (or cluster centre), beginning with some values named seed points. The distances between input data points and centroids are determined and inputs to the closest centroid are allocated K-means [4].

Steps for k-means Clustering used by Chitraa et al. [4]:

1. We achieve several transactions once the data are cleaned and the sessions are recognised. Each transition involves multiple URLs.
2. The number of input neurons to the ANN are most likely to be the pages of the website.
3. Alpha value is computed which is the threshold value, i.e. the resemblance between two transactions.
4. Choose any one transaction amongst all the recognised transactions as the centroid.
5. Select another transaction and calculate the distance between the centroid and this transaction, if the distance computed is less than alpha, it can be concluded that the second transaction is a separate cluster altogether.

6.3   Fuzzy Logic in Web Mining

Fuzzy logic can be considered a generalised form of classical logic. Lotfi Zadeh (mid-1960s) attempted to develop fuzzy logic, which was intended to assist the calculations and predictions of problems that required the use of inaccurate data or the formulation of inference rules generically using the diffusion classification.

The unit interval [0,1] is the most popular range of membership function values that are usually taken into consideration. Let μ_A represent the membership function of the fuzzy set A, which can be expressed as $\mu_A:U \to [0,1]$, so that for every $x \in U$, $\mu_A (x) = \alpha, 0 \le \alpha \le 1$.

Fuzzy logic consists of a multiple-valued logic, in which the values lie within the range 0 to 1, which is considered to be "fuzzy", to express the reasoning that occurs in humans. These values are termed the truth values for certain logical propositions. A particular proposition X may have 0.6 as the truth value and its complement

0.5. As per the negation operator that is applied, these two truth values do not compulsorily have to sum up to 1. Fuzzy logic can be applied as a model to interpret the NNs and their characteristics. They can be useful in determining the network specifications without the need to apply a learning algorithm [17]. Fuzzy logic is used in cases where the inaccuracy and ambiguity levels are high and to handle the variations on a more continual level. The degree of truth is the value possessed by the propositions as a mathematical model for imprecision and vagueness. Another important quantity is the membership function, i.e. a function that describes and relates the degree of membership to a value in the domain of sets in the fuzzy set.

A real-life instance may be as follows: old air conditioners used to function using an on–off system. The interpretation of each one may be denoted by a specific fuzzy set. Let an AC be at 27 °C. When the temperature rises above 25 °C, the unit is turned on, whereas if the temperature drops below 20 °C, then it might be interpreted as the AC being turned off. Fuzzy rules such as "the cooling power would be turned up a little, in case the ambient air gets warmer; the cooling power would be turned down moderately if the temperature is dropped" would be applied. The application of the aforementioned concept would simplify the functioning of the machine and would result in a more steady and consistent temperature.

The reason for using fuzzy clustering for Web mining as opposed to the traditional clustering is because the web data possess fuzzy traits and properties. The researchers have determined the way in which soft computing techniques (such as NNs, fuzzy logic, GAs etc.) can be applied to web mining as a tool to improve the efficiency of retrieval or processing of the results obtained from the search

6.4 Ant Colony Optimisation (ACO) in Web Mining

Ant colony optimisation (ACO) is one of the algorithms used for the intention of examining social insects such as ants or bees as opposed to imitating human intelligence (Fig. 5).

The manner in which ants optimise their trail looking for food by releasing pheromones, a chemical substance produced and released into the environment affecting the behaviour or physiology of other ants on the trail is what works as a motivation behind this meta-heuristic. They, in fact, tend to traverse the path with the shortest length, which they determine with the help of the strength of the pheromones' smell on the paths visited by the other ants, which is usually the shortest, as the shortest path takes the least amount of time to traverse. Therefore, even if the same number of ants traverse both the long and the short path, the number of traversals would be higher in the short paths depositing more pheromones on the way, which would make the other ants choose the shorter path. This intelligent behaviour arising from these otherwise unintelligent creatures has been called "swarm intelligence".

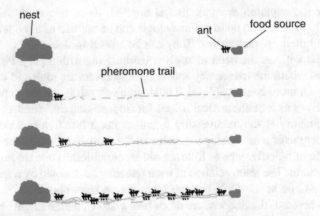

Fig. 5 Pheromone trails in natural ant colonies

Algorithm 1

Init *pheromone* τ_{ij};
repeat for all ants i: *construct_solution(i)*;
    for all ants i: *global_pheromone_update(i)*;
    for all ants edges: evaporate *pheromone*;
$$(\tau_{i-j} = (1 - \rho)\tau_{i-j})$$

construct_solution(i) :
init *ant*;
while not yet a solution **do**
    expand the solution by one edge probabilistically according to the *pheromone*;
$$(\tau_{\rho i-j} / \textstyle\sum_{\rho i-j^*} \cdot \tau_{\rho i-j^*});$$
end while

global_pheromone_update(i) :

for all edges in the solution **do**
    increase the *pheromone* according to the quality;
$$(\Delta \tau_{i-j} = 1/length\ of\ the\ path\ stored)$$
end for

Fig. 6 Pseudo-code for ant colony optimisation (ACO)

This elementary concept is to the web user a chain of the web pages visited, or sessions. The ants, which have been created artificially, are made to undergo training with the help of a web session clustering technique to refashion a preference vector of the text that depicts the priorities of the users to choose the preferred group of keywords. Moreover, the behavioural pattern of browsing in the future is predicted by these ants (Fig. 6).

6.5 Particle Swarm Optimisation

Particle swarm optimisation (PSO) is a method of computation that solves a problem and obtains the most efficient solution by frequently attempting to enhance a candidate solution pertaining to the provided estimate of a quality. It considers a population of candidate solutions, the particles in this case, and provides the result with the help of the motion of the particles in the search-space by the location and velocity of the particle determined by obtaining the values using the formulas mathematically. Every particle's motion is affected by its position, which is best known locally, although it is mentored in the direction of the most preferred positions in the search-space, which are renewed as and when preferable positions are determined by other particles. This is what is intended to provide the optimal solution for the swarm.

It is derived from the idea of how a flock of birds or a school of fish behaves socially and is preferable for various applications owing to the lower number of parameters to be considered for adjustments.

7 Hybrid Techniques for Sentiment Analysis

Sometimes, two or more traditional data mining methods can be combined to yield better performance while examining social media data. Such composite techniques consisting of various traditional techniques are commonly referred to as hybrid techniques. It is often found that the hybrid techniques have an edge over their traditional counterparts as they exploit the advantages of the individual techniques to provide better accuracy and results. Some of these hybrid techniques are discussed in depth below.

7.1 Analysis of SVM-PSO Hybrid Technique [2]

The messages or certain remarks on Twitter can be used to review a movie with the help of opinion mining or sentiment analysis by applying text mining techniques, natural language processing and computational linguistics to categorise the movie into good or bad based on message opinion. In this application, the focus is on the binary categorisation, which groups the data into the positive and negative classes, with positive depicting a good message/review and negative depicting a bad one, for a particular movie that is being analysed. It uses tenfold cross-validation and a confusion matrix for the authentication process, and the precision level of the support vector machine forms the basis of the rationale. To advance the selection of the most appropriate parameter to tackle the dual optimisation problem, hybrid particle swarm optimisation (PSO) has been chosen.

The pre-existing datasets for Twitter sentiment messages are very rare. The set of data used for the research was accumulated by Hsu et al. [7]. In that, the messages selected for the training data had emojis, which were discarded. The test data were manually obtained and contained 108 positive and 75 negative tweets from the user, which were noted manually. The dataset was then preprocessed and cleansed before being used for the research.

The result depicts the enhancement of the extent of precision from 71.87% to 77%.

One of the tiers of the flow chart (Fig. 7) is machine learning and classification, which involves the SVM. The below figure articulates the functioning of the SVM. The first step would be indexing the term of opinion ascending. Next, all the terms

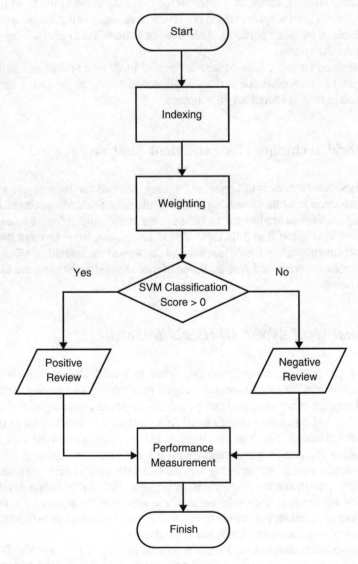

Fig. 7 Flow of the support vector machine model

are weighted according to their features. If the weighting score is more than zero (weight >0), the term is categorised as a positive review. If it is not, then the term is categorised as a negative review.

Figure 8 below illustrates the formation of PSO with population size, inaction weight, and generations without improvement. This is followed by the examination

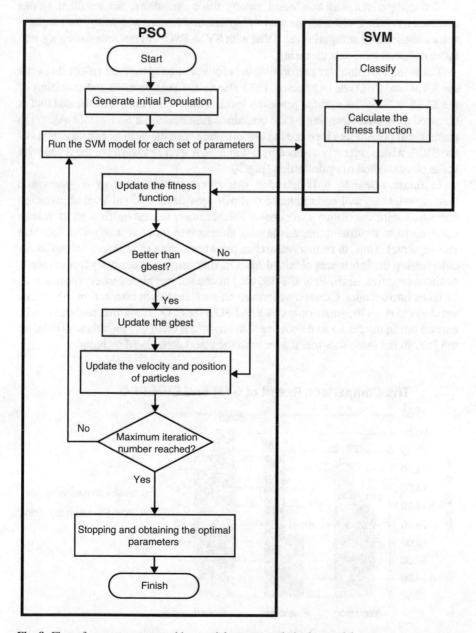

Fig. 8 Flow of support vector machine particle swarm optimisation model

of each particle's fitness. Next, is the comparison of the fitness functions and the locally best and globally best parameters. After that is completed, the velocity and position of each particle is updated until the convergence of values of the fitness task. Once convergence is finished, the SVM classifier is fed with the global best particle in the swarm for training. Lastly, the SVM classifier is trained.

The comparison was conducted among three quantities, the resultant values obtained from the SVM with the use of N-grams and feature weighting, the resultant values obtained by comparing the SVM with SVM-PSO without data cleansing, and those obtained with data cleansing.

The results obtained by performing the aforementioned method reflect that after the SVM and PSO are hybridised, PSO affects the performance and precision of the SVM. 77% is the optimal precision level, which is suitable for this, and that is obtained after data cleansing by the combined functioning of SVM-PSO. Although there is still potential to increase the performance with the help of enhancements of the SVM, which is mostly assisted by combining it with or varying of the SVM with some other method of optimisation (Fig. 9).

In future research, it is intended that more combinations of N-grams and feature weighting will be introduced that will provide an optimal level of precision compared with the above experiment. Furthermore, the categorisation of results obtained from sentiment analysis is only categorised into two categories (positive and negative). Thus, in further research, a class consisting of multiple categories for categorising the inferences obtained from further sentiment analyses (for example, positive, negative, neutral/not affected etc.) instead of just binary categorisation may be taken into account. Currently, there are no such studies in connection with social networks to show the superiority of SVM-PSO over PSO. Therefore, research can be carried out in the future to determine if the hybrid SVM-PSO technique dominates the PSO in the same way that it does with the standalone SVM technique.

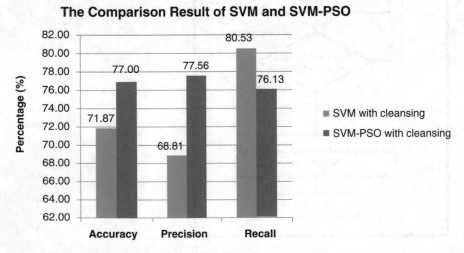

Fig. 9 The comparison result of the SVM and SVM-PSO after data cleansing

7.2   Analysis of SVM-ACO Hybrid Technique [11]

Currently, Twitter is considered one of the most prevalent micro-blogs for making short and frequent posts, which equips Twitter users to comment on a particular post, be it reviews on a movie or a book, as "tweets". These can be helpful in determining the general outlook of the public and opting for better marketing strategies. Like the previous analysis, the Tweets obtained from the Twitter data can be categorised into two sets: positive and negative, only this time with the approach of support vector machine (SVM), a ML algorithm and ant colony optimisation (ACO) hybrid strategy. The precision on an average of this categorisation increases from 75.54% to 86.74% with the application of SVM and SVM-ACO hybrid respectively.

The flow chart given in Fig. 10 depicts the procedure of opinion mining of Tweets from Twitter data. The tiers of the diagram are as follows:

1. Accumulation of data: the application programming interface (API) of Twitter offers the provision of obtaining Tweets with the keywords concerned with the help of a programmatic technique.
2. Preliminary processing: the preliminary processing of the Twitter data is done to determine and remove the redundant data elements from the input dataset. In this way, it performs data cleaning before the actual data processing to improve the precision of the categorisation. As the data obtained from the public may subsume the elements, which may not affect the polarity of the reviews, if these data are considered for calculations, it would lead to increased complications of the process of categorisation. The measures of preliminary processing involve erasing punctuation characters, such as the comma, period, apostrophe, quotation marks etc. It also incorporates a number filter to refine the numerical terms, a case converter to transform all text forms to lower case etc.
3. Generation of features: features distribute the text obtained from the source into a positive or negative category. Some of the popular methods of weighing the features include term frequency (TF), which is used for the generation of the feature vector and term frequency-inverse document frequency (TF-IDF), which are calculated with the help of Eqs. (1) and (2).

$$\text{TF} = \frac{\text{Number of times term } t \text{ appears in a document}}{\text{Total number of terms in the document}} \tag{1}$$

$$\text{IDF} = \log_e \frac{\text{Total number of documents}}{\text{Number of documents with term } t \text{ in it}} \tag{2}$$

4. Categorisation by applying SVM: A hyper-plane is generated by the SVM for categorisation (Fig. 11). The hyper-plane can maximise the nearest training instance of both categories i.e. maximise the functional margin. The prime goal is to reduce the error by generalisation and does not allow over-fitting to affect it.
5. Categorisation by applying the SVM-ACO hybrid: One of the concepts that applies swarm intelligence to tackle problems would be ACO. A trail of synchronously moving ants works coherently to deal with a sub-problem that is a

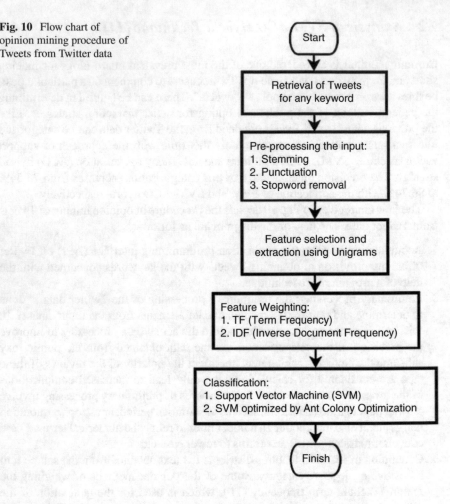

Fig. 10 Flow chart of opinion mining procedure of Tweets from Twitter data

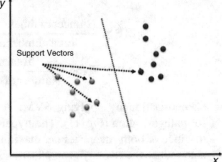

Fig. 11 Hyper-plane that segregates the two classes in the SVM: positive and negative

part of the prime problem and provides clarified solutions to this problem. Each ant builds its own solution to the problem incrementally after each movement. An ant updates its trail value according to the components used in its solution after evaluating and completing its solution or at its construction phase. The search for future ants is directed and affected to a large extent by this pheromone value (Fig. 12).

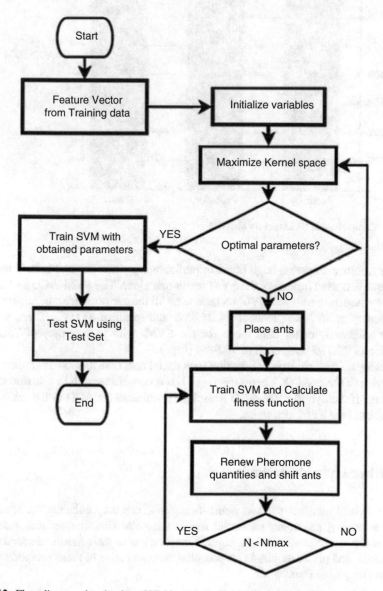

Fig. 12 Flow diagram showing how SVM is optimised with the help of ACO

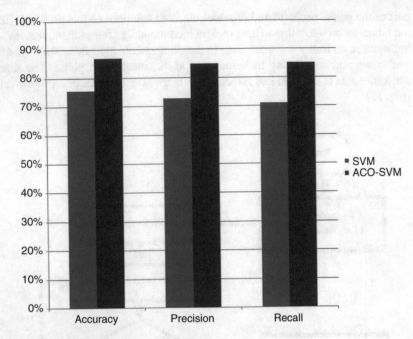

Fig. 13 Comparison of SVM and SVM-ACO

The results exhibited a huge boost in performance when the SVM-ACO hybrid technique was used instead of the SVM technique alone. The SVM-ACO technique outperformed the traditional SVM technique on all the performance parameters with an accuracy of 86.74%, precision of 84.85% and recall of 85.05%. These values were significantly better than those for the SVM, with an accuracy of 75.54%, precision of 72.77% and recall of 70.98% (Fig. 13).

There is no such study in connection with social networks to show the superiority of SVM-ACO over ACO. Therefore, there is a scope of research in this domain to determine if the hybrid SVM-ACO technique dominates the ACO in the same way it dominated the SVM technique.

8 Other Hybrid Methods

Certain hybrid methods such as neuro-fuzzy methods may enhance the speed of the web mining procedure on social media datasets. This chapter has analysed the existing techniques, methods, algorithms along with their merits, demerits and limitations, and places emphasis on potential enhancements of these methods using the soft computing framework.

8.1 Neuro-Fuzzy Hybrid Technique

Neuro-fuzzy is a term used in the field of artificial intelligence in the context of fusion of artificial neural networks and fuzzy logic. Proposed by J.S.R. Jang, the hybridisation of neural networks and fuzzy logic creates an intelligent system that coalesces and merges the reasoning of the fuzzy logic systems with the connectionism of the artificial neural network, resulting in a linguistic model that is is much more adept than both fuzzy logic (FL) and the ANN.

In Shivaprasad et al. [18], various soft computing techniques such as fuzzy logic, neural networks, genetic algorithms (GA) and ant colony optimisation (ACO) have been analysed by the authors in an attempt to evaluate their efficacy in mining the data from the social media. The research concluded that there is a requirement for fast hybrid techniques for the analysis of social media data. Hybridisation techniques such as neuro-fuzzy hybridisation can render better solutions at a reduced cost and at a faster speed. Neuro-fuzzy hybridisation is ideally suited as it has the advantages of both fuzzy logic and neural networks. It is adept in handling unclean data and in modelling the non-linear decision boundaries.

Shivaprasad et al. [18] have proposed an ingenuous hybrid neuro-fuzzy model for the mining of web data. After the collection, pre-processing and cleaning of web data, the clustering of data is carried out using the Fuzzy C Means clustering algorithm. Each cluster is unique as it comprises users exhibiting similar browsing behaviour. The artificial neural networks are constructed specific to the goals and objectives to be accomplished while imitating the architecture and information representation patterns of the human brain. In the learning process, each pattern is present in the input nodes and associated with output nodes with differential weights. The iterative process followed adjusts and re-assigns the weights between the input and the output nodes until a predetermined termination condition is met. The training of the neural network is carried out by giving the input of the clustering algorithm as the input of the neural network, whereas the output of the clustering algorithm is the target output of the neural network. The performance of the system is measured using the mean square error (MSE), which is the average of the square of error between the outputs of the neural network and the target outputs from the clustering algorithm. A multi-layered forward neural network is used which is trained with back-propagation learning algorithms.

9 Conclusion

The prevalence of social media usage has enhanced the inquisitiveness in the field of opinion mining. This has made it an essential task for several corporate institutions and companies to categorise the sentiments from the texts and public reviews in an efficacious manner. Keeping this in mind, we have tried to analyse the hybrid techniques such as neuro-fuzzy, SVM-ACO and SVM-PSO after briefly explaining the traditional techniques and giving a brief description of all the important concepts and methodologies that are referred to in this chapter. Towards the end of each hybrid intelligent technique, the results and relevant inferences pertaining to that particular technique are taken into account.

It can be clearly observed that hybrid intelligent techniques are more efficient than their individual components. The experiments indicate that the usage of hybridised techniques significantly increases the precision level percentage. Moreover, the precision level of the hybrid methods can be further increased with the help of improvements in the existing version of the components used to generate the respective hybrid models. It could be achieved by using it in another combined arrangement or modification of the existing component models with some other method of optimisation. We have mentioned how a particular technique is suitable for a certain type of data-set and how new hybrid intelligent techniques can be created to produce better optimisation and accuracy.

10   Future Scope

- The results of various hybrid methods are examined experimentally and observationally on datasets of varying sizes for the purposes of opinion mining. Even though the hybrid intelligent techniques are more efficacious, there is a requirement and scope for development of new techniques that would be more efficient and accurate.
- In the future research on the various hybrid methods, an extended analysis is expected to be carried out to examine how further enhancement can be made in each respective area and how several domains and region-specific parameters influence the sentiment analysis results. Expanding the concept of opinion mining to various other domains may culminate in innovative results.
- A broader variety in combining the N-grams and assigning weight to the attributes and features that provide more precision than the existing ones can be researched and taken into account.
- The current techniques usually classify the data into two categories (binary classes) that are either favourable or unfavourable (or positive and negative). In further research, on opinion classification of more than these two categories (like an additional neutral category) is expected to be included in the results and analysis of hybrid intelligent techniques.
- There is a huge chunk of data from the health sector that can be analysed using hybrid techniques for better prediction of diseases. In future, the sentiments of the crowd can be analysed and relevant conclusions can be drawn regarding the winning party in the elections.
- Apart from social media data, the use of hybrid techniques in the medical sector also has huge potential. In their research, Mishra et al. [14] have shown that SVM-ACO can be successfully used to find the best subset descriptors for the classification of anti-hepatitis peptides.

- Furthermore, it would be expected that the various other hybrid techniques, apart from the conventional combinations, would be researched and implemented after their feasibility and analysis had been weighed up.

We consider that this chapter may also be used to obtain novel ideas for new lines of research or to continue the lines of research proposed here.

References

1. Aggarwal, C., Zhai, C.: A survey of text clustering algorithms. In: Mining Text Data, pp. 77–128 (2012). doi:10.1007/978-1-4614-3223-4_4
2. Basari, A, Hussin, B., Ananta, I., Zeniarja, J.: Opinion mining of movie review using hybrid method of support vector machine and particle swarm optimization. Proc. Eng. **53**, 453–462 (2013). doi:10.1016/j.proeng.2013.02.059
3. Berry, M.: Automatic discovery of similar words. In: Survey of Text Mining: Clustering, Classification and Retrieval, pp. 24–43. Springer, New York (2004)
4. Chitraa, V., Selvadoss Thanamani, A.: An enhanced clustering technique for web usage mining. Int. J. Eng. Res. Technol. **1**(4), 5 (2012)
5. Feitosa, R., Labidi, S., dos Santos, A.S.: Hybrid model for information filtering in location based social networks using text mining. In: 2014 14th International Conference on Hybrid Intelligent Systems (2014). doi:10.1109/his.2014.7086206
6. He, W., Zha, S., Li, L.: Social media competitive analysis and text mining: a case study in the pizza industry. Int. J. Inf. Manage. **33**, 464–472 (2013). doi:10.1016/j.ijinfomgt.2013.01.001
7. Hsu, C.-Y., Yang, C.-H., Chen, Y.-C., Tsai, M.-C.: A PSO- SVM lips recognition method based on active basis model. In: 2010 Fourth International Conference on Genetic and Evolutionary Computing (2010). doi:10.1109/icgec.2010.188
8. Jusoh, S., Alfawareh, H.: Agent-based knowledge mining architecture. In: International Conference on Computer Engineering and Applications, pp. 526–530. IACSIT Press, Singapore (2009)
9. Jusoh, S., Alfawareh, H.: Techniques, applications and challenging issue in text mining. Int. J. Comput. Sci. Issues **9**, 431–436 (2012)
10. Kamruzzaman, S., Haider, F.: A hybrid learning algorithm for text classification. In: International Conference on Electrical & Computer Engineering, 1st edn. (2004)
11. Kaur, J., Sehra, S., Sehra, S.: Sentiment analysis of twitter data using hybrid method of support vector machine and ant colony optimization. Int. J. Comput. Sci. Inf. Secur. **14**, 222–225 (2016)
12. Khan, F., Bashir, S., Qamar, U.: TOM: Twitter opinion mining framework using hybrid classification scheme. Decis. Support Syst. **57**, 245–257 (2014). doi:10.1016/j.dss.2013.09.004
13. Kumar, R., Sharma, M.: Advanced neuro-fuzzy approach for social media mining methods using cloud. Int. J. Comput. Appl. **137**, 56–58 (2016). doi:10.5120/ijca2016908927
14. Mishra, G., Ananth, V., Shelke, K., et al.: Classification of anti hepatitis peptides using Support Vector Machine with hybrid Ant Colony Optimization. Bioinformation **12**, 12–14 (2016). doi:10.6026/97320630012012
15. Mitra, S., Pal, S., Mitra, P.: Data mining in soft computing framework: a survey. IEEE Trans. Neural Netw. **13**, 3–14 (2002). doi:10.1109/72.977258
16. Mohana, R., Umamaheshwari, K., Karthiga, R.: Sentiment classification based on latent dirichlet allocation. In: International Conference on Innovations in Computing Techniques (2015)
17. Rojas, R.: The backpropagation algorithm. In: Neural Networks, 1st edn., pp. 149–182. Springer, Berlin (1996)

18. Shivaprasad, G., Reddy, N., Acharya, U., Aithal, P.: Neuro-fuzzy based hybrid model for web usage mining. Proc. Comput. Sci. **54**, 327–334 (2015). doi:10.1016/j.procs.2015.06.038
19. Sumathy, K.L., Chidambaram, M.: Text mining: concepts, applications, tools and issues an overview. Int. J. Comput. Appl. **80**, 29–32 (2013). doi:10.5120/13851-1685
20. Verma, P., Keswani, N.: Web usage mining: identification of trends followed by the user through neural network. Int. J. Inf. Comput. Technol. **3**, 617–624 (2013)
21. Zafarani, R., Abbasi, M., Liu, H.: Social Media Mining: An Introduction, 1st edn. Cambridge University Press, Cambridge (2014)
22. Zhai, C., Massung, S.: Text Data Management and Analysis: A Practical Introduction to Information Retrieval and Text Mining, 1st edn. ACM Books, New York (2016)

Social Network Metrics Integration into Fuzzy Expert System and Bayesian Network for Better Data Science Solution Performance

Goran Klepac, Robert Kopal, and Leo Mršić

Abstract Basic parameters for social network analysis comprise social network common metrics. There are numerous social network metrics. During the data analysis stage, the analyst combines different metrics to search for interesting patterns. This process can be exhaustive with regard to the numerous potential combinations and how we can combine different metrics. In addition, other, non-network measures can be observed together with social network metrics. This chapter illustrates the proposed methodology for fraud detection systems in the insurance industry, where the fuzzy expert system and the Bayesian network was the basis for an analytical platform, and social network metrics were used as part of the solution to improve performance. The solution developed shows the importance of integrated social network metrics as a contribution towards better accuracy in fraud detection. This chapter describes a case study with a description of the phases of the process, from data preparation, attribute selection, model development to predictive power evaluation. As a result, from the empirical result, it is evident that the use of social network metrics within Bayesian networks and fuzzy expert systems significantly increases the predictive power of the model.

Keywords Bayesian networks • Fuzzy expert system • Social network analysis • Social network metrics

1 Background

Big data, as a paradigm show social network analysis (SNA) to be one of the dominant methodologies for finding patterns within network data. Network data can be found all around us in virtual spaces present through social networks and

G. Klepac (✉)
Raiffeisen Bank Austria, Petrinjska 59, Zagreb, Croatia
e-mail: goran@goranklepac.com

R. Kopal • L. Mršić
IN2Data, Marohnićeva 1/1, Zagreb, Croatia
e-mail: robert.kopal@in2data.eu; leo.mrsic@in2data.eu

© Springer International Publishing AG 2017
H. Banati et al. (eds.), *Hybrid Intelligence for Social Networks*,
DOI 10.1007/978-3-319-65139-2_2

within companies, in which clients and users have interactions or as known with mobile communication in telecommunication companies. Social network metrics as a part of SNA plays an important role in knowledge discovery within social network data. Finding out information about influencers within the network is not an easy task, even when sophisticated network metrics exist. With a history of almost 100 years, SNA has developed under many influences from various fields, primarily sociology and mathematics, but its implementation as an interdisciplinary technique ranges from physics and computer science to communication science, economics, and even psychology. Social network analysis as an academic discipline was introduced in the UK in 1920–1930, in the research field of anthropology. The term social network was first coined by anthropologist Roger Brown, who suggested that the social structure is similar to a network and that interpersonal communication among individuals resembles the relationship between one node and another nesting in the network [29, 32]. At the turn of the twentieth century, Georg Simmel was the first scholar to think directly in terms of social networks. He advanced the idea that society consists of a network of patterned interactions, and argued that society is merely the name for a number of individuals, connected by interaction. All human behavior is the behavior of individuals, but Simmel suggested that much of it can be explained with regard to the individual's group affiliation, and constraints imposed upon him by particular forms of interaction [7, 30]. The visualization of networks and the early sociograms began with a social psychologist of Romanian origins, Jacob Levy Moreno, with his sociometry method, which was created in the 1930s. This method introduced the systematic analysis of social interaction in small groups, and as such paved the way for quantitative analysis in social network approaches [26]. Sociometry had a crucial influence on sociology and related fields. Over time, its influence has significantly diminished. However, "a complete understanding of sociometry provides tremendously powerful structures and tools for use not only in small group interactions but also wherever and whenever interpersonal dynamics come into play" [28].

The period between the 1940s and 1960s is considered by some to be "the Dark Ages" in the history of the development of SNA. In this period, there was no generally recognized approach to social research that included the "structural paradigm." Freeman argues that SNA was still "not identifiable either as a theoretical perspective or as an approach to data collection and analysis." After that, there was a period of a rapid growth in the number of scholars who focused their work on combining different tracks and traditions in studying social networks. One large group gathered around Harrison White and his students at Harvard University: Ivan Chase, Bonnie Erickson, Harriet Friedmann, Mark Granovetter, Nancy Howell, Joel Levine, Nicholas Mullins, John Padgett, Michael Schwartz, and Barry Wellman. The so-called renaissance of SNA spread among Harvard scholars, resulting in a number of important articles that influenced social scientists everywhere, regardless of their field [13, 23]. The study of social networks develops in parallel on two tracks, sociological (networks) and mathematical (graphs), and consequently, SNA is based on two different approaches. Apart from the aforementioned sociological aspect with a small-world phenomenon, an equally important aspect for the development of SNA is the mathematical one, namely, combinatorics, or more precisely, graph theory. Graph theory is a mathematical tool for depicting and explaining social networks. The mathematical foundation for SNA came from Hungarian mathematicians Paul Erdos and Alfred Renyi, with their model of generating random graphs, or in terms of SNA, random networks. A few years before Milgram's small-world experiment, in 1959, Erdös and Rényi suggested the first algorithm for random graph generation, which uses the same probability of choosing node pairs from a given set of nodes, and links them based on the previously defined probability. According to this model, the set that has n nodes forms every connection with a given probability of p, and the formation is independent across links [10–12]. The Erdos-Renyi model is one of the oldest and most intensely studied random network models, whose basic feature is the possibility of mathematically calculating most of its average properties. This model has an intuitive structure and was a starting point for most of the later models. As it presumes that connections in a network are the result of a completely random process, properties of random networks may help to obtain insights into properties of some other networks, social or economic, for example. Some of these properties, intensively studied today, are the ways of distribution of connections over nodes, ways in which networks are connected in terms of their capability of finding ways from one node to another, the average and maximum path lengths, and the number of isolated nodes. Therefore, random networks can be a useful benchmark for comparing observed networks because these comparisons help to identify those elements of social structure that are not the result of mere randomness, but must rather be assigned to other factors. Two key points that differentiate this model from real-world networks are low network clustering and unrealistic degree distribution. Over the past 15 years, from the late 1990s onwards, the research into networks has diffused over a number of scientific fields through the study of complex networks. Graph theory in its original form, designed by Erdos and Renyi, their model of random graphs, and some other mathematical models of networks that have been studied in the past, lacked certain properties that occur in real networks, such as a heavy tail in the degree distribution, a high clustering coefficient, assortativity or disassortativity among vertices, community structure, and hierarchical structure. However, the study of these nontrivial topological features have brought together researchers from many areas in an attempt discover patterns of connection and other specific structural features of real networks that are neither purely random nor purely regular in their nature. Today, graph theory is regularly used for finding communities in networks. The number of algorithms for detecting communities grows rapidly daily, which enables routine real and business problems to be resolved using this methodology. By detecting small networks, communities, and subgraphs, and their structures, hierarchical, circular or other, it is possible to apply SNA to various problems, e.g., cross-selling in marketing, or fraud and money laundering in revenue insurance.

2 Social Network Metrics

2.1 Social Network Metrics Basics

Generally speaking, a social network is a social structure made up of actors (individuals or organizations) called "nodes," which are tied (connected) by one or more specific types of interdependency, such as friendship, kinship, common interest, financial exchange, dislike, or relationships of beliefs, knowledge or prestige. In other words, SNA views social relationships in terms of network theory consisting of nodes and ties (also called edges, links, or connections). Nodes are the individual actors within the networks, and ties are the relationships between the actors. In SNA, graphs and matrices are typically used to represent social network data. In the network structure modelled as a graph, vertices represent actors, and edges represent ties, which show the existence of a relation between two actors. Apart from indicating the links between vertices, the data also provide us with additional information on directionality and link weights [27]. The network can also be used to measure social capital, the value that an individual obtains from the social network. Social network analysis is equally used in mathematical sociology and quantitative analysis. The major difference in approach is the fact that statistical analysis methods need a probabilistic approach to calculate probability distributions of relationship strengths. Mathematical approaches to the analysis of networks use structural descriptive (deterministic) methods, assuming that the measured relationships and their strengths represent the real or final status of the network. A crucial advantage of statistic models according to Wasserman and Faust is that they "can cope easily with some lack-of-fit of a model to data," whereas descriptive or deterministic models "cannot be relaxed in this way." "Deterministic models usually force the aspect of social structure of interest (such as reciprocity, or complete transitivity or structural equivalence) to be present in the model, while statistical models assume these aspects to be absent" [6]. Centrality is a key concept in SNA. This measure gives a rough indication of the social power of a node based on how well they "connect" the network. "Betweenness," "closeness," and "degree" are all measures of centrality. A highly centralized network is dominated by one object who controls information flow and may become a single point of communication failure. A less centralized network has no single point of failure; therefore, people can still pass on information even if some communication channels are blocked. SNA software allows us to calculate betweenness, closeness and degree centrality measures to provide different perspectives on the social relationships within the network. The centrality measures can also be influenced by taking into account the direction of links and the weightings applied to them.

Betweenness centrality is based on the importance of intermediaries in the communication network. By asking ourselves "How crucial is an object to the transmission of information through a network? How many flows of information are disrupted or must make longer detours if a person stops passing on information or disappears from the network? To what extent may a person control the flow of information owing to his or her position in the communication network?" [6], we can determine the extent of the importance of an actor in the network. Betweenness centrality stands for the extent to which a node lies between other nodes in the network. It measures the number of paths that pass through each entity. It takes into account the connectivity of the node's neighbors, giving a higher value for nodes that bridge clusters. The measure reflects the number of people with whom an object is connecting indirectly through their direct links. In this sense, betweenness centrality measures the control that an individual has over the interaction of others who are not directly connected. This may identify entities with the ability to control information flow between different parts of the network. These are called gatekeeper entities.

Gatekeepers may have many paths running through them, allowing them to channel information to most of the others in the network. Alternatively, they may have a few shorter paths running through them, but still play a powerful communication role if they exist between network clusters.

Link betweenness centrality measures the number of paths that pass through each link. This can help to identify key connections of influence within the network. A link through which many paths pass may be a significant route for information exchange between entities. In the following figure, the circled link has the highest measure of betweenness because it provides a key path through which information may flow between various parts of the network.

Closeness centrality is based on the reachability of a person within a network: "How easily can information reach a person?" It measures the proximity of an entity to the other entities in the social network (directly or indirectly). Thus, closeness is the inverse of the sum of the shortest distances between each individual and every other person in the network. The closeness centrality of a vertex is higher if the total distance to all other vertices is shorter. An entity with a high measure of closeness centrality has the shortest paths to the other entities, allowing them to pass on and receive communications more quickly than anybody else in the organization. Information has to travel much further to and from an entity on the edge of a network that is attached only to few other entities; therefore, they will have a lower measure of closeness centrality. One of the most common and important approaches to indexing the distances between actors is the geodesic. The geodesic distance or diameter between pairs of actors is the measure of closeness used to describe the minimum distance between actors. In simplest terms, geodesic is the shortest path between two vertices. However, the geodesic distance describes only a single connection between a pair of actors (or, in some cases several actors), and sometimes we are not interested in the shortest connection, but rather in the sum of all connections between actors. In directed networks, when measuring the diameter, the direction of the edge needs to be considered [21].

The average geodesic distance is the measure that explains the closeness in the network, i.e., the average distance value between the entities. If the value is high, it means that a small number of entities have close contacts and that there is a large number of steps/entities between them. On the other hand, if the value is low, entities in the network are close. This measure reminds us of Stanley Milgram and his "six degrees of separation" [25].

Degree centrality is a basic measurement of how well connected an entity is by counting the number of direct links each entity has with others in the network. In undirected networks, entities have only one degree value, whereas in directed networks, each vertex can have three values for each entity: Outdegree, whose number depends on the number of links going from the entity Indegree, whose number depends on the number of links coming to the entity Degree or total degree, standing for the sum of the outdegree and indegree numbers of links. This can reveal how much activity is going on and who are its most active members. This measure can be defined as a certain social capital of the entity, and the entity with the highest number of direct links can be considered a central entity. Link degree. Directed links can be taken into consideration when calculating centrality measures. A link with arrows added to it represents the directed flow of information between entities, either in a single direction or in both directions. This may have an important bearing on how quickly information is passed from one part of the network to another. For example, a person may receive information from many others in the network, but only send information to a select few. The centrality measures for an entity through which information is channeled in both directions are higher than the measures for an entity through which information is channeled one way.

A somewhat more sophisticated measure of centrality is the measure of the importance of a node in a network called eigenvector. By singling out the entity with the highest score, eigenvector makes it possible to identify who in the network mobilizes other entities and indirectly controls the network. It assigns relative scores to all nodes in the network based on the principle that connections to nodes having a high score contribute more to the score of the node in question. Google's PageRank is a variant of the eigenvector centrality measure.

Ranking deals with the notions prestige and ranking through acyclic decomposition. In a directed graph, prestige is the term used to describe a node's centrality. Degree prestige, proximity prestige, and status prestige are all measures of prestige. In social networks, there are many techniques for calculating the so-called structural prestige of a person. It is important to distinguish between structural prestige and social prestige: the first depends on the data from which we are able to infer a structure so that the structural prestige of a person only reflects his or her social prestige. The indegree of a vertex represents the popularity of the person represented by the vertex. To measure popularity, we need to have a directed network.

Domains represent extended prestige to indirect in-links so that the overall structure of the network is taken into account. The input domain of a vertex is defined as the number or percentage of all other vertices that are connected by a path to this vertex. A restricted input domain restricts the maximum path length. In well-connected networks, it is recommended to limit the input domain to direct neighbors (i.e., to use popularity instead) or those at a predefined maximum distance [3]. Once we have determined the nature of our network, we can start discovering the clusters and the hierarchy. The technique used for extracting discrete ranks from social relations and for determining the hierarchy, is called acyclic decomposition. Authority centrality and hub centrality measures were first introduced by Kleinberg [16] for information retrieval purposes in directed (and acyclic) networks. These metrics are suitable for simultaneously measuring mutually reinforcing centrality arising from nodes pointing to other nodes (hubs), and from nodes being pointed to by other nodes (authorities). These metrics are very useful in studying the possibilities of the diffusion of information inside the network. When analyzing the diffusion process, it is critical to determine the number of nodes that can be connected, and, even more importantly, to identify the nodes that have better chances of diffusing information through the social structure using their connections. To be able to identify such nodes, we first need to recognize the nodes within the network that have a high first degree of centrality with a good sequence of nodes in their connection paths.

2.2 Social Network Metrics Hybrid Approach

Social metrics measures are powerful measures with which we are in a position to discover hidden relations between nodes. Each network metric plays a specific role in discovering relations, and in important node recognition. During the SNA stage, analysts are in a position to combine different metrics with the intention of revealing hidden knowledge [1, 2, 4]. Usage of a social metrics combination is not always an easy task, especially in a situation where we would like to create filters for network structures based on social network metrics [9]. Interpretation of such a defined social network metrics combination could be hard and may ensure periodical analysis on the same filters. This chapter gives a solution illustrated through the use of a fuzzy expert system and Bayesian networks, which, along with other attributes, incorporate social network metrics in hybrid structures, helping to discover more complex social network measures as a combination of standard ones. This approach gives standardization through concepts such as fuzzy logic and Bayesian networks, in addition to transparency, which is frequently applicable in different time periods on some social network data sets with comparison capabilities. Fuzzy logic and fuzzy expert systems are knowledge-oriented systems that can use social network metrics as linguistic variables. Such created linguistic variables can be used through rule block definition for finding complex relationships within networks [20, 22]. Bayesian networks are conditional probability-oriented models, which, along with a statistical approach, take into consideration expert knowledge as well. Through Bayesian network construction it is important to discover mutual influences between basic network metrics for sensitive and predictive Bayesian network creation. The aforementioned approach could be used not only for network node filtering, but also as a base for predictive modeling and network segmentation. It may also be interesting as an island in the sea of social network methodology exploration.

3 Improving the Power of the Use of SNA Metrics

3.1 Role of Fuzzy Logic

A fuzzy set is an extension of a classical set. If X is the set of values and its elements are denoted by x, then a fuzzy set A in X is defined as a set of ordered pairs by the following formula [14, 15, 24, 31].

$$A = \{x, \mu_A(x) | x \in X\};$$ (1)

Membership function can be defined for betweenness centrality, eigenvector, closeness, degree, and similar measures. Each variable can be expressed as a linguistic variable and by the definition visible in Fig. 1.

The definition of linguistic variable may include social network metrics. Definition of social network metrics through linguistic variables can be integrated (combined) through fuzzy rule blocks. In this type of situation, social network metrics such as betweenness centrality, eigenvector, closeness, degree, and other similar measures can be structured in new categories that describe some of the new characteristics. Network influence can be defined as a combination of specific social network metrics characteristic of the problem space that we would like to solve.

Table 1 shows the methodology of the definition of rule blocks based on social network metrics.

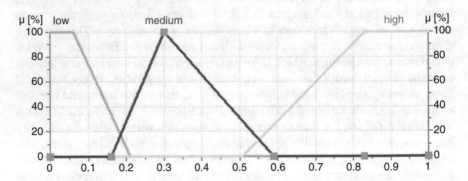

Fig. 1 Linguistic variable definition on social metrics

Table 1 Definition of fuzzy rules on social network metrics and an example

Rule number	IF eigenvector	AND closeness	THEN influence
Rule 1	Low	Low	Low
Rule 2	Medium	Low	Low
Rule 3	High	Medium	Medium
Rule 4	Low	High	Medium

As is visible from the rule block created, it is possible to combine social network metrics into complex structures within one rule block. Connected rule blocks make a fuzzy expert system. A fuzzy expert system can contain linguistic variables based on social network metrics in addition to nonsocial network variables. A fuzzy expert system provides the opportunity for a complex influence calculation constructed from different social network metrics. Such a constructed expert system is periodically usable, which is interesting with regard to the observation of social metrics trends, even though its construction is complex and includes many of them.

3.2 Role of Bayesian Networks

As mentioned above, using social network metrics in situations when large numbers of them are being used and combining them can be difficult for interpretation and in projects with empirical data. Bayesian networks can integrate numerous social network metrics to calculate the output from the network structure. The basic idea is for example to create a predictive model that has social network metrics as predictors along with other variables to make a much more accurate prediction. For Bayesian network construction, it is important to carry out an attribute relevance analysis to find out which attribute has predictive power. Also, it is not only a question of influence on an observed target variable, because it is important to realise the mutual influence among disposable variables that are used for Bayesian network construction. In that process, along with attribute relevance analysis, expert knowledge plays an important role. Basically, from the perspective of predictive modeling, there are two main types of predictive models from the perspective of the target variable.

- Binomial target variable models.
- Multinomial target variable models.

In the case of predictive models with a binomial target variable, the common approach to attribute relevance analysis is the use of weight of evidence and information value calculation. In situations with a multinomial target variable, which is much more complicated, an information gain calculation can be used. Information gain can be calculated using the following formula [21]:

$$\text{Info}(D) = -\sum_{i=1}^{n} p_i \log_2(p_i) \tag{2}$$

where p_i is the probability that an arbitrary tuple in D belongs to class C_i. This measure is recommended for use in situations where the output variable has more than two states for prediction. There are many measures that can be used for this purpose. Information gain is presented as one possible solution for attribute relevance analysis in situations when we are operating with more than two states in an output variable.

Formally, it can be expressed as [18]:

$$Pr[P_j] = [\text{node parents}] \tag{3}$$

Or more precisely as [18]:

$$Pr[P_j] = \prod_{i=1}^{n} P_r[P_j \mid Y_i, \ldots, Y_n] \tag{4}$$

Where:

$Pr[P_j]$ represents probability for j-th instance, which can also be a social metric
Y_i represents probability of i-th characteristics within a Bayesian network expressed as a state value of a Bayesian network node, which can also be a social metric

In the case of building predictive models, which uses social network metrics as well, variables can be observed through the time perspective and can be expressed as a temporal node within a defined Bayesian network. It means that recognized nontemporal variables in the process of attribute relevance analysis can be connected with temporal elements (temporal nodes). In this case, we are talking about dynamic Bayesian networks.

For temporal process modeling we should extend the definition of Bayesian network to include the kth-order Markov process. For this purpose, we should define the transition model as [19]:

$$p(Z_t \mid Z_{t-1}, Z_{t-2}, \ldots, Zt - k) = \prod_{i=1}^{N} P(Z_t^i \mid Pa(Z_t^i)) \tag{5}$$

where Z_t^i is the i-th node at time t, and $Pa(Z_t^i)$ are the parents of Z_t^i.

If we are talking about Bayesian networks, or dynamic Bayesian networks, we should be aware that Bayesian networks operate with categorical values.

If we are using Bayesian networks, it does not mean that the whole model is concentrated only on target variables. It is important to calculate attribute relevance analysis for each variable within the model where the observed variables became target variables and connected variables became predictors. This methodology ensures model predictability and sensitivity. In that process, social network metrics can act as target variables and predictors. Depending on the variable type, we are applying information values or information gain calculations.

In this regard, we can build a model that integrates different social network metrics with an influence on each other and on other nonsocial network metric variables. Their mutual effects should improve model reliability and accuracy.

4 Applying the Model to Empirical Data

The proposed methodology combines the usage of social network metrics data with other types of data within the concept such as Bayesian networks and fuzzy logic. To illustrate this, the proposed methodology is shown for fraud detection purposed on insurance company data. The traditional approach may also take into consideration a variety of social network metrics in finding solutions in an unstructured and unsystematic way. The proposed methodology provides a structured and systematic way of finding patterns within social networks when seeking influencers and suspicious activities. The structured method takes into account a complex definition of influence within social networks, in addition to a structured method of periodic observation of defined influences through a combination of traditional social metrics. Bayesian networks unite expert knowledge and the statistical approach to network construction, and fuzzy expert systems lean only on expert knowledge and human expertise.

Many tools for detecting insurance fraud target specific verticals of business (such as receivables management) and then build a framework system around it. To make the system more robust, the methodology requires a broader view that includes integration of all potential areas related to fraud (claims, premiums, employees, vendors, suppliers) in a wider perspective.

An optimal fraud model is multidimensional, involving strategic, organizational, process, and technology and data components. In terms of strategy, it should successfully reduce the financial and operational impacts of fraud; senior leadership must set the right tone from the top. Furthermore, this means conveying the seriousness of fraud as a business issue that needs top managers need to be aware of. As anti-fraud strategies and plans take shape, organizations need to change boundaries to ensure that the key functions are working collaboratively to address the problem. With regard to organizational resources, there is a clear need for skilled resources and adequately defined roles, responsibilities, and accountability. Multiple layers to be considered in the model such as underwriters, claims, adjusters, screeners, and investigators all working toward common fraud detection and prevention goals. Implementing a fraud model may necessitate some adjustments inside the organization, or at least better coordination across functional lines, with possible special investigative units that are optimally integrated with their counterparts in claims or other areas. In respect of processes, the detection stream is the key component of the anti-fraud model. There are plenty of inputs for the company to consider while preparing the model.

4.1 Problem Description

The described methodology has been applied to the empirical data of a Croatian insurance company regarding car insurance products. The task was to find a predictive model for fraud detection development. Existing models based on the fuzzy expert system and Bayesian networks contained variables with no social metric variables. Their predictive power was relatively high, as shown in the following table, especially because in the last 2 years, there had been an increasing trend toward suspicious cases being detected, which increased the fraud rate within the portfolio. Fraud detection is always a challenging task regarding the nature of the fraud, and this is why the company wanted to try some other approaches, techniques, data sources, and concepts to improve existing fraud detection models. Initial SNA showed some nodes with strong influences within the network, which represents the whole process of the insurance lifetime cycle from contracting to the end of the insurance process. Some of the cases that were recognized as suspicious or that proved fraud showed significant dependence on nodes recognized as having a significant influence within the network. Nodes within social networks could be participants in accidents, the car, location, and other similar attributes. The task was to prove or reject the hypothesis that nodes with a great influence might be recognized as some of the elements of the insurance process and that payment of damages by insurance might be greatly influenced in fraudulent cases.

Proving this hypothesis could mean that there is a chain of people involved who are participants in fraud. Including social metrics in the Bayesian network after attribute relevance analysis, and including social metrics in the fuzzy expert system should increase predictive power. Most important is exploratory analysis of potentially suspicious participants (external or internal) connected with fraudulent cases. Increasing concentrations, rising influence, the significant role of an object in the observation of fraudulent cases could raise questions about the role of that object in the insurance process.

In a car accident, drivers and passengers usually exchange addresses and phone numbers, and then submit them to the insurer. However, one address or one of the vehicles may already be included in one or more of the reported adverse events. The ability to obtain this information may save a lot of time for the insurer and provide him with new information about the case and/or indicate potential fraud. SNA allows the insurer to create a proactive view of the links between participants and previously identified patterns of dishonorable conduct. For this purpose, traditional methods and SNA are combined with other tools and methods. Such a hybrid approach involves organizational business rules, statistical methods, the analysis of links, the analysis of behavior patterns, and SNA of large data sets to gain more obvious recognized forms of connectivity. Typical examples of publicly available data that may be included in the production of hybrid models are the judgment of the courts, enforcement, criminal records, and the frequency of address change. A hybrid insurance model can comprehensively rank the claims based on the assigned rating. If the assessment of the claim shows, for example, an address that has been included in the records of a previous fraud or a vehicle that has been involved in other accidents that exceeds a specified limit, automated detection of such cases may make it difficult for investigators to focus on a specific suspicious claim. The role of SNA is to recognize influencers within a social network. This information became part of the Bayesian network and the fuzzy expert model. The idea is to increase the predictive power of the Bayesian network and the fuzzy expert model using SNA metrics.

4.2   SNA Social Metrics Calculation

For social metrics network calculation, an Anaconda environment with the NetworkX library in Python programming language was used on transactional and preprocessed data. Data were captured from Oracle databases and stored in local data structures for analytical purposes. The following code was used for social network metrics calculation.

```
import networkx as nx
from operator import itemgetter

G=nx.Graph()
G=nx.read_pajek("Fraud.net")

print "Info:"
print nx.info(G)

print "Degree histogram:"
print nx.degree_histogram(G)

print "Density:"
print nx.density(G)

print "Number of nodes:"
print G.number_of_nodes()

print "Number of edges:"
print G.number_of_edges()

dc= nx.degree_centrality(G)
Sorted_degree = sorted(dc.items(), key=itemgetter(1),
    reverse=True)
print "Sorted degree:"
print Sorted_degree [0:5]

bc= nx.betweenness_centrality(G)
Sorted_betweenness = sorted(bc.items(), key=itemgetter(1),
    reverse=True)
print "Sorted betweenness:"
print Sorted_betweenness [0:5]

cc= nx.closeness_centrality(G)
Sorted_closeness = sorted(cc.items(), key=itemgetter(1),
    reverse=True)
print "Sorted closeness:"
print Sorted_closeness [0:5]
```

Calculation results were used in a Bayesian network and a fuzzy expert system. The Bayesian network was developed in GeniE (http://www.bayesfusion.com/) and the fuzzy expert system was developed in FuzzyTech (www.fuzzytech.com). Social network analysis had the task of investigating and discovering potential concentrations among participants in the contracting process for cases that were discovered as fraudulent. SNA also had the task of investigating potential links between elements within cases that were discovered to be fraudulent and other participants who were not directly (obviously) connected with that case. For that purpose, a meta-model with all participants, internal and external, and with the objects of the insurance, was created.

4.3   Usage of Bayesian Networks with SNA Metrics

Bayesian networks are commonly used in risk modeling, but they are not a purely statistical predictive model (like, for example, neural networks or logistic regression) because their structure can also depend on expert knowledge. The Bayesian network structure could be settled in an algorithmic way, but from the business perspective or the perspective of model efficiency and overall performance, it is recommended that the Bayesian network structure be modified by expert knowledge [5, 8, 17].

Expert involvement in a network structure may not be a guarantee that the network will be optimal, but it is aligned with business perception of the problem space.

The developed model of Bayesian network had 18 nodes, five of which contained the social network metrics eigenvector, betweenness, indegree, outdegree, and closeness. Connectivity between nodes has been recognized by the use of information gain between nodes, in combination with the necessary path condition (NPC) algorithm. The aim variable is fraud and for social network metrics with other nonsocial metric attributes, information gain was calculated. In a first step, attribute relevance analysis was calculated with regard to the aim variable fraud. The next step was attribute relevance analysis in such a way that all variables recognized as being important as predictors for fraud target variable became target variables. For each of these variables, information gain was calculated, and important attributes regarding the new target variable were connected within the Bayesian network. In a final stage, the NPC algorithm was applied to that structure to find additional links.

Information gain has discovered mutual connections between nodes. As is visible from Fig. 2, those connections are discovered between nonsocial metric nodes, and between social metric nodes. Variables declared as node $1 \ldots n$ are nonsocial metric nodes. After attribute relevance analysis, it was evident that social metrics play an important role within the model. Social metrics gave additional predictive power to previously developed Bayesian networks, revealing hidden patterns and participants, which are important in fraudulent processes. Even this model was built with the intention of suspicious score calculation. With social network metrics, it gives the

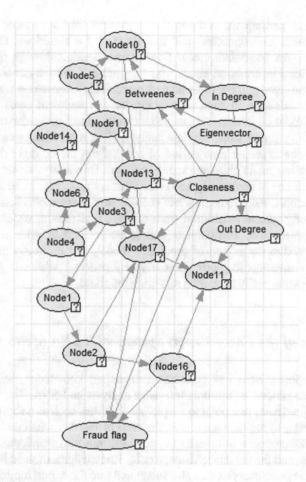

Fig. 2 Bayesian network with social network metrics

power for an exploratory analysis that can directly discover nodes (participants) into a fraudulent process. Nonsocial metric nodes are not declared with attribute names to protect company privacy. As a result of SNA, there was a significant connection between some of the contractors and suspicious accidents and between some of the lawyers and suspicious accidents. This fact leads to the conclusion that there is a potential doubt regarding the fraudulent activities if requests for paid compensation are connected to a specific lawyer. Regarding contractors and a significant link between some of the contractors and suspicious accidents, it was discovered that there are only three unique contractor IDs within the system for contracting in a company, and this was why SNA discovered the connection between contractors and suspicious accidents.

Knowledge revealed from social network metrics gives a detailed insight into what often happens in a fraudulent process. Social network metrics reveal the fact that some of the participants in the insurance process are lawyers and contractors, that a few others are connected, and that there was a significant concentration of connections for fraudulent cases. Deeper investigation through the Bayesian network provided information about who they are and who was connected with whom in a significant number of fraudulent cases. This information was invaluable for deeper investigation of each case or cluster of fraudulent cases for detailed investigation about other characteristics that can be useful for understanding fraud from a business point of view. Survival analysis (Cox regression) was used for the evaluation of the expected time period of fraud, from the contract date for each suspicious participant in the contracting process. This analysis was the starting point for deficiency in contracting and other business processes in relation to those insurance products. Additional analysis showed some extreme factors such as the short period of payment realization from damage registration, which leads us to additional investigation. The point is that social network metrics can be useful in many ways, directly as part of predictive model, and indirectly as milestones for additional investigation, providing new information, which is important to understanding what really happened regarding the fraud.

4.4 Usage of the Fuzzy Expert System with SNA Metrics

The fuzzy expert system was built based on expert knowledge, and discovered relations in the process of attribute relevance analysis and SNA. The fuzzy expert system contained rules that point to suspicious activities such as a short time period from contracting to the damage report, the time of the day when the damage was entered into the company information system, the number of accidents in a certain period of time, and the insurance/damage ratio. Knowledge extracted in the process of attribute relevance analysis was also integrated into the expert model, in addition to social network metrics. As a result, the expert fraud model classified each contract into the following categories:

- Unsuspicious
- Less suspicious
- Suspicious
- Very suspicious

First processing from the expert fraud model recognized 0.09% very suspicious cases, 0.23% suspicious cases, and 1.5% less suspicious cases. Some of the dominant mutual characteristics of all suspicious cases were the short time period from contracting to damage report, the high number of accidents in the last 2 years, the significant connection between some of the lawyers and suspicious accidents. All suspicious cases were investigated individually, and the model proved its efficiency after the investigation.

Social network metrics within the fuzzy expert model had the task of investigating and discovering potential concentrations among participants in the contracting

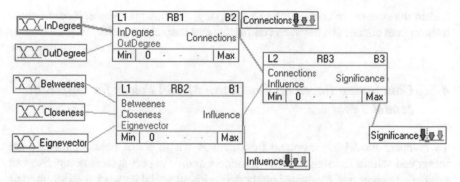

Fig. 3 Part of the fuzzy expert system, which includes social network metrics

process for cases that were discovered to be fraudulent. Also, SNA had the task of investigating potential links between elements within cases that were discovered to be fraudulent and other participants not directly (obviously) connected with that case. The fuzzy expert system structure and rules and linguistic variables were created with the help of experts. During the interviewing stage, linguistic variables, their connections, and rules were constructed. Elements of social network metrics within the fuzzy expert system are shown in Fig. 3.

In the existing fuzzy expert system, additional rule blocks and social network oriented linguistic variables have been added, as is visible from figure. Indegree and outdegree were added through the independent rule block, and other measures such as eigenvector, betweenness, and those that are mostly "influence" oriented were added through another rule block. These two independent rule blocks were connected through a mutual rule block and a mutual output variable. A predictive model operates with few variables and depends on a statistical sample. Expert models can contain many variables, and a statistical sample is not relevant for expert model development. This block construction was plugged into the existing expert system for fraud detection, and was connected with other variables and rule blocks. This maneuver played the role of expert model accuracy and reliability improvement. Also, fuzzy expert systems introduce new dimensions of views on data, with accurate measurement on fraud influence regarding values of social network metrics. Fuzzy expert system confirmed findings from Bayesian networks and additionally showed new hidden behavioral patterns, which included social network metrics.

As a result, from SNA, there was a significant connection between some of the contractors and suspicious accidents and between some of the lawyers and suspicious accidents. High social network metrics (such as betweenness) regarding some fraudulent case types and specific layers can be an indication of potential fraud. This fact leads to the conclusion that there is potential doubt about fraudulent activities if the request for paid compensation is connected to a specific lawyer. Regarding contractors and the significant link between some of the contractors and suspicious accidents it was discovered that there are only three unique contractor IDs within the system for contracting in a company, and this was why SNA discovered a significant connection between contractors and suspicious accidents.

4.5 Conclusion Based on Predictive Model Power for Different Model Types

To compare model improvement in situations where social network metrics are integrated within existing predictive models and/or expert system comparison of predictive power was conducted on models without social network metrics, models with social network metrics, and combined models with social network metrics, as is shown in Table 2.

A test was performed on data samples that proved or did not prove, but very suspicious cases were marked as fraudulent. Data samples were constructed from 7-year history data. On such a data sample, the fraud rate was 1.8%. Some cases with similar actors were repeated in that period. Repeated fraud cases and very suspicious cases contributed to a fraud rate of 0.75%. Table proves the positive influence of the inclusion of social network metrics on existing models. Even if it does not look spectacular regarding area under curve numbers, they are not negligible values. Fraud activities do not have frequent appearance, and each increase in predictive power increases the model's reliability. On the other hand, one fraudulent case can result in a massive financial loss to the company, and each successful detection contributes to the prevention of a large financial loss. The best results are visible in situations in which Bayesian network are combined with fuzzy expert systems, where both models contain social network metrics. It proves that the hybrid model approach has its own values. In our case, we made hybrid models using social network metrics, in addition to more than one concept. Fraud detection is a very challenging area, which does not suffer from a lack of creativity on either side: fraudsters and fraud detection modelers. Social network metrics in such types of fraud detection models, as is visible from the results, can contribute to much more sophisticated models.

Table 2 Model type and area under the curve

Model type	AUC
Expert system	76.56
Bayesian network	64.57
Expert system with SNA metrics	77.12
Bayesian network with SNA metrics	64.93
Bayesian network and fuzzy expert system with SNA metrics	77.83

5 Conclusion

The case studies presented show the importance of the integration of different analysis concepts based on traditional and big data sources for efficient fraud detection solution. Each methodology or concept does not provide answers to all questions, but when integrated, they create a synergy effect that can lead to a much more efficient solution than in situations in which only one methodology or concept is used. The case study showed the importance of result chaining through different systems, and this can result in some new pattern or knowledge revelation. The traditional approach to fraud detection, which leans on reporting and predictive modeling, is not sufficient, especially in the era of big data environments and new techniques suitable for pattern recognition from big data sets. The case study presented shows one of the possible solutions for fraud detection in insurance, which may be applicable in complex business environments, and use the advantages of advanced hybrid analytical models. The solution unites different concepts instead of using a single concept and model. The advantage of this approach is a model that can detect fraud with much more efficiency. Fraud, by nature, is a non-frequent appearance, which complicates predictive statistical model development. Another issue is that even in a case where hypothetically, fraud is frequent in a way that for modeling purposes we have significant data samples, it is unrealistic to expect uniform pattern recognition, which is the basis for predictive modelling. However, we should not neglect the potentials of fraud predictive modeling, but nor should we lean only on that concept as a universal concept for fraud detection. The solution is development of a fraud system that unites different paradigms such as predictive modeling, expert knowledge integration via fuzzy expert systems, and SNA. The presented approach is a good tool for discovering fraudsters' creativity. Leaning on one concept or method often leads to inadequate results, because, as has already been mentioned, fraudsters are very creative and if a pattern becomes obvious, it is often too late to apply data science for finding solutions, because a company may have serious problems. If internal fraud prevention is based merely on the predictive model, then a significant part of potentially fraudulent behavior remains uncovered, because predictive models operate on only six to twelve variables, which is insufficient to cover all possible fraudulent situations. To be much more efficient, fuzzy expert systems are introduced to cover as far as possible potential gaps and indicators, which are not frequent but significant for fraud detection.

In addition to developing a predictive model, parts of the process should be used to understand relations and key drivers. Predictive models should not be used merely as a probabilistic calculator because in that case, the business background remains unknown and hidden. An understanding of the business background, in this case the cause of fraud, is a crucial factor for model efficiency and strategic management.

References

1. Abraham, A., Hassanien, A-E., Snášel, V.: Computational Social Network Analysis Trends, Tools and Research Advances. Springer, London (2010)
2. Aharony, N., Pan, W., Cory, I., Khayal, I., Pentland, A.: Social fMRI: Investigating and Shaping Social Mechanisms in the Real World, Pervasive Mob. Comput. **7**(6), 643–659 (2011)
3. Akoglu, L., Vaz de Melo, P.O.S., Faloutsos, C.: Quantifying Reciprocity in Large Weighted Communication Networks, PAKDD 2. Lecture Notes in Computer Science, vol. 7302, pp. 85–96. Springer, Berlin, Heidelberg (2012)
4. Altshuler, Y., Pan, W., Pentland, A.: Trends Prediction Using Social Diffusion Models, Social Computing, Behavioral-Cultural Modeling and Prediction. Lecture Notes in Computer Science Series, pp. 97–104. Springer, Berlin, Heidelberg (2012)
5. Bishop, C.M.: Pattern Recognition and Machine Learning. Springer, New York (2006)
6. Carrington, P.J., Scott, J., Wasserman, S. (eds.): Models and Methods in Social Network Analysis, pp. 248–249. Cambridge University Press, Cambridge (2005)
7. Coser, L.A.: Masters of Sociological Thought: Ideas in Historical and Social Context, 2nd edn. Harcourt, New York, NY (1977)
8. D'Agostini, G.D.: Bayesian Reasoning in Data Analysis: A Critical Introduction. World Scientific, New York (2003)
9. Easley, D., Kleinberg, J.: Networks, Crowds, and Markets: Reasoning about a Highly Connected World. Cambridge University Press, Cambridge (2010)
10. Erdös, P., Rényi, A.: On random graphs. Publ. Math. (Debrecen) **6**, 290–297 (1959)
11. Erdös, P., Rényi, A.: On the evolution of random graphs. In: Publication of the Mathematical Institute of the Hungarian Academy of Sciences, vol. 5 (1960)
12. Erdös, P., Rényi, A.: On the strength of connectedness of a random graph. Acta Math. Acad. Sci. Hung. **12**, 18–29 (1961)
13. Freeman, L.C.: The Development of Social Network Analysis: A Study in the Sociology of Science. Empirical Press, Vancouver, BC (2004)
14. Fuller, R., Carlsson, C.: Fuzzy Reasoning in Decision Making and Optimization. Physica-Verlag, Heidelberg (2002)
15. Hampel, R., Wagenknecht, M., Chaker, N. (eds.): Fuzzy Control: Theory and Practice. Physica, Heidelberg
16. Jackson, M.O.: Social and Economic Networks. Princeton University Press, Princeton, NJ (2010)
17. Jaynes, E.T.: Probability Theory. The Logic of Science. Cambridge University Press, Cambridge (2003)
18. Jensen, F., Nielsen, T.: Bayesian Networks and Decision Graphs. Springer, New York (2007)
19. Kjarluff, U., Madsen, A.: Bayesian Networks and Influence Diagrams: A Guide to Construction and Analysis. Springer, New York (2013)
20. Klepac, G.: Discovering behavioural patterns within customer population by using temporal data subsets. In: Bhattacharyya, S., Banerjee, P., Majumdar, D., Dutta, P. (eds.) Handbook of Research on Advanced Hybrid Intelligent Techniques and Applications, pp. 321–348. IGI Global, Hershey, PA (2016)
21. Klepac, G., Kopal, R., Mršić, L.: Developing Churn Models Using Data Mining Techniques and Social Network Analysis, pp. 1–308. IGI Global, Hershey, PA (2015). doi:10.4018/978-1-4666-6288-9
22. Klepac, G., Kopal, R., Mršić, L.: REFII model as a base for data mining techniques hybridization with purpose of time series pattern recognition. In: Bhattacharyya, S., Dutta, P., Chakraborty, S. (eds.) Hybrid Soft Computing Approaches, pp. 237–270. Springer, New York (2015)
23. Lauritzen, S.L., Nilsson, D.: Representing and solving decision problems with limited information. Manage. Sci. **47**, 1238–1251 (2001)
24. Leonides, C.: Fuzzy Logic and Expert Systems Applications. Academic, New York (1998)

25. Milgram, S.: The small-world problem. Psychol. Today **1**(1), 61–67 (1967)
26. Moreno, J.L.: Sociometry, Experimental Method, and the Science of Society. Beacon House, Ambler, PA (1951)
27. Pinheiro, C.A.R.: Social Network Analysis in Telecommunications. Wiley, Hoboken, NJ (2011)
28. Remer, R.: Chaos theory links to Morenean theory: a synergistic relationship. J. Group Psychother. Psychodrama Sociom. **59**, 38–45 (2006)
29. Scott, J.: Social Network Analysis: A Handbook. Sage Publications, London (1987)
30. Simmel, G.: How is society possible? In: Levine, D. (ed.) On Individuality and Social Forms. University of Chicago Press, Chicago, IL (1908/1971)
31. Zadeh, L.A., Kacprzyk, J. (eds.): Fuzzy Logic for the Management of Uncertainty. Wiley, New York (1992)
32. Zhang, M.: Social network analysis: history, concepts, and research. In: Fuhrt, B. (ed.) Handbook of Social Network Technologies and Applications, pp. 3–22. Springer, New York, NY (2010)

Community Detection Using Nature Inspired Algorithm

Sunita Chand and Shikha Mehta

Abstract Community detection in social networks has become a dominating topic of the current data research as it has a direct implication in many different areas of importance, whether social network, citation network, traffic network, metabolic network, protein-protein network or web graph etc. Mining meaningful communities in a real-world network is a hard problem owing to the dynamic nature of these networks. The existing algorithms for community detection depend chiefly on the network topologies and are not effective for such sparse graphs. Thus, there is a great need to optimize those algorithms. Evolutionary algorithms have emerged as a promising solution for optimized approximation of hard problems. We propose to optimize the community detection method based on modularity and normalized mutual information (NMI) using the latest grey wolf optimization algorithm. The results demonstrate the effectiveness of the algorithms, when compared with other contemporary evolutionary algorithms.

Keywords Community detection • Grey wolf optimization • Modularity • NMI

1 Introduction

Humans have always preferred to live in society by making groups to form small or big communities. The communities could evolve based on race, colour, caste, religion, gender, job category, interest, hobbies etc. Nowadays, with the rapid advancement in the speed of the Internet and its reduced cost, we try to socialize through social networks, despite being separated by long distances across the globe. This has been made possible because of the social networks available to us. These networks have become the backbone of virtual socialization and help people to interact and share their latest updates with their friends and relatives. Thus, social

S. Chand (✉)
Hansraj College, University of Delhi, New Delhi, India
e-mail: sunitamk@gmail.com

S. Mehta
JIIT Noida, Jaypee University, Noida, India
e-mail: shikha.mehta@jiit.ac.in

© Springer International Publishing AG 2017
H. Banati et al. (eds.), *Hybrid Intelligence for Social Networks*,
DOI 10.1007/978-3-319-65139-2_3

47

networks such as Facebook, Twitter, MySpace and LinkedIn have occupied a major share of internet usage. These social networks have not only helped us to be friends with others, but have become an important tool for propagating ideas and our views on a large scale. These social networks have emerged as an important tool for politicians for their mass campaigning during the election period. It has also helped retailers and businesses to promote their products. At the same time, ordinary people have opted for social media to raise their voices against any social crime or injustice. Considering the importance of social networks, there is a great need to analyze these networks from various aspects. Social network analysis may be done at different levels, i.e. at a node level, at a group level or at the network level [7]. We may need to identify the group or cluster to which a node belongs. Sometimes we may also need to analyze the social network to find the most prominent personality on the network. The one with the maximum number of followers across all the communities may give us an indication of how important a person is in the group. Analogous to our social physical groups or communities, the virtual network of nodes on social media may also be categorized into communities. Recognizing these communities could be of particular interest to researchers and analysts. For instance, finding the community of a collaboration network may help us to discover groups of researchers working in the similar domain or having expertise in a similar field. In social networks, community detection helps by recommending a person to be a friend of a person who may belong to a certain group, e.g. a group of ex-school friends, a group of office colleagues, a family group, or a group of close friends. We may also recommend a person to join a particular group based on the links that exist between that person and a member of that group. Hence, investigating community has become a hot topic of research these days. The analysis may be based on several parameters, e.g. degree centrality, betweenness, closeness, information and rank. Community detection (CD) in a social network has come forth as a fundamental tool for social network analysis that lays the foundation for several higher-level analytics. It is associated with partitioning a social network into various cohesive subgroups or clusters in such a way that there is a strong cohesion or interconnection among the nodes of a subgroup and weak cohesion among the nodes of different subgroups.

The process of detecting communities within large networks, if automated, could prove to be very helpful for the network analysts. For instance, communities in a web graph may represent a group of web sites concerned with related topics, whereas communities in an electronic circuit or a biochemical network may correspond to some sort of functional unit in that network [9]. Social networks may be defined as vast networks of vertices representing individuals or entities, and the edges connecting the vertices as the interaction between them. They can be represented by a graph $G(Vn, Em)$ where V is the set consisting of n nodes and E is the set of m edges connecting them. Practically, in social networks, "n" and "m" keep changing dynamically, leading to increased complexity. The value of "n" may be in hundreds, thousands or even millions, as in Facebook, Twitter etc. Graph problems such as clique, the independent set problem, vertex colouring, edge colouring, travelling salesman problem, graph isomorphism and graph partitioning are considered to be NP hard problems. As social networks can also be represented as graphs, analysis of social networks is also considered to be a nondeterministic polynomial time (NP) hard problem [9]. With the rapid industrialization and technological transformation, there is a great need for optimization in every aspect of problem solving. As the number of dimensions, size of problems and the number of variables increase, the problems become complex and difficult to optimize. Social networks analysis is one such problem where there is a great need for optimization. In this chapter, we attempt to present the computer algorithms for the optimized extraction of communities from social network data. We present an introduction to social networks and the techniques for social network analysis in Sect. 2 of this chapter. Section 3 concentrates on nature-inspired algorithms and their applications. Section 4 gives a detailed insight into community detection in social networks and various solutions to community detection. We present a novel approach to community detection in Sect. 5 of this chapter. Sect. 6 concludes this chapter.

2 Social Networks and Their Analysis

The network has been defined as a set of entities or nodes that are linked by edges connecting them. These nodes may be used to represent anyone or anything. An edge connecting any two nodes represents the relation or the link existing between those nodes [2]. These edges can have characteristic properties such as direction, which signifies the flow of information, and the weight, signifying the cost of the link. The cost may be the distance or the financial cost. Examples of networks include cellular phone networks [33], biological networks, protein-protein interaction networks, communication and transportation networks, collaboration networks, the Internet, the world-wide-web and social networks [15].

2.1 Social Networks

Social networking services, also known as social networks or social media, provide an online platform that facilitates people building social relationships or social networks of friends.

The nodes in social networks represent the people and the edges signify the type of social interaction existing between them. For instance, the relationship may be the common personal interest, friendship, religion, activities, backgrounds, career, real-life connections etc. These sites are aimed at connecting individuals and creating clusters. The social networks may be categorized into three types:

1. **Socializing social networks:** These networks are primarily used for socializing with friends and family members (for example, Facebook and MySpace).
2. **Networking social networks:** These are used primarily for non-social inter-personal interaction and have a specific target audience (e.g. LinkedIn (Business)).

3. **Social navigation network services:** These sites are aimed at helping users to find specific resources and information (e.g. Jeevansathi.com for match-making, Goodreads for books, Last.M for music, Match.com for dating, Yahoo personals, HASTAC, Ning, TermWiki for education and PatientsLikeMe for healthcare).

With the rapid increase in social networking, other applications of social media are also emerging, such as:

1. **Real-time web:** Based on the pattern of live radio and television broadcast, these social media wherein the users can broadcast what they are doing (Twitter, Clixtr and Facebook with "Live feed")
2. **Location based web:** Merges business solutions and technologies, e.g. cloud computing with social networking concept (e.g. Foursqures, Gowalla, Yelp etc).

In recent years, traffic from social networks has occupied a huge share of the overall traffic on the Internet.

2.2   Representation of Networks

Like other networks, a social network may be mathematically modelled as a graph $G = (V, E)$, where the set V represents the set of individuals and the set E represents the set of edges. The total number of individuals can be represented by $n = |V|$ and the total number of edges in the network by $m = |E|$. A community in a network is a subgroup of the vertices that have a high density of edge interaction within them, and a relatively lower density of edge interaction between them. This problem of dividing the network into k unknown communities in a network, may be formulated as partitioning the graph into k partitions that have dense intra-connection but sparse inter-connections. As a network is represented as a graph, we represent the network consisting of N nodes using a $N * N$ adjacency matrix A, where

$$A(i,j) = \begin{cases} 1 & \text{if node } i \text{ is connected to node } j, \\ x & \text{if node } i \text{ is not connected to node } j. \end{cases}$$

Network Terminology

Let us familiarize ourselves with the key terminology of a network [35]:

Actors/Agents/ Point/Node: Social entities such as individuals, cities or organizations.
Arc/Tie/Edge/Link/Line: It represents interactions or relationships among actors.
Dyads: Represented by a pair of actors and the relationship between them.
Triad: A subgroup of three actors and the relationships among them.
Subgroup: Consists of nodes and the relations existing among them.
Group: Set of all actors under observation, on which ties are to be measured.
Relation: A set of a specific kind of ties among members of a group.
Social network: A finite set or sets of actors defined with the relationship or relationships among them.

2.3 Social Network Analysis

Social network analysis (SNA) may be defined as the measuring and mapping of relationships and flows among computers, URLs, people, groups, organizations and other connected knowledge/information entities [40].

2.3.1 Levels of Analysis

Generally, network methods are appropriate at certain levels of analysis [14, 35], which are discussed below:

Actor Level network is analyzed at actor level. The scope of this analysis is the properties and the associated methods pertaining just to actors, e.g. finding the "prominence" of an actor within a group, as measured by metrics such as centrality and prestige, actor-level expansiveness and popularity parameters embedded in stochastic models and metrics for individual roles such as bridges, isolates and liaisons.

Dyadic Level The type of analysis that is applicable to pairs of nodes and the ties between them. It includes finding the distance and reachability between any two nodes, structural and any other notions of equivalence, and probabilistic tendencies towards correlativity and the statistical model-based dyadic analyses for the modelling of various states of a dyad.

Triadic Level This is based on theoretical statements about the balance and transitivity and hypothesizes the behaviours for groups of three actors and their interconnections.

Subset Level This type of analysis helps researchers to study and find subsets of actors that are homogenous with respect to some properties of the network, e.g., cliques and cohesive subgroups containing actors who are "close" to each other, finding subsets of actors having similar behaviour with respect to certain model parameters from some stochastic models and determining the positions of actors through positional analysis etc.

Network Level Last, there are metrics and methods that work on entire groups and all interconnections. It includes graph theoretical measures such as connectedness and diameter, group level measures of centralization, density and prestige. In addition, this type of analysis also focuses on block models and role algebras.

2.3.2 Metrics (Measures) in Social Network Analysis

We need to understand some important properties of social networks, e.g. size, density, degree, reachability, geodesic distance and diameter. These properties may be used in various forms of social network analysis. Hence, we provide a brief discussion of these properties, as mentioned in [32]:

Centrality This metric gives an approximation of the social power of a node based on degree of its connectivity to other nodes in the network. "degree", "betweenness" and "closeness" are all measures of centrality.

Betweenness Centrality This centrality measure reflects the number of individuals with whom a person is indirectly connecting through their direct links. A vertex occurring on many shortest paths between other vertices has higher betweenness than those that do not.

Closeness The extent to which an individual is close to all other persons in a network (either directly or indirectly). It may be considered a measure of the time taken by the information to spread from a vertex to other vertices in the network, that are reachable from the given vertex. The shortest distance between two nodes may also be known as the "geodesic distance".

Degree A count of the number of edges incident upon a given vertex gives the measure of degree centrality. It often reflects the immediate risk of node for catching any information or virus, flowing through the network. If the network is directed, then degree centrality can be measured by two separate measures; namely, indegree and outdegree. Indegree represents the count of the number of edges coming to a vertex, whereas outdegree tells us the number of edges going out of a given vertex to other nodes in the network.

Bridge A bridge is an edge that connects two vertices in different components of the graph. The number of components increases on deleting this edge from the network. It is also known as cut-edge or cut arc or an isthmus. An edge may be a bridge only if it is not contained in any cycle.

Prestige The prestige of an actor may be found by the number of connections entering to it. In other words, it indicates the node's centrality. Different measures of prestige include "Degree prestige", "status prestige" and "proximity Prestige". The prestige of an actor increases as it participates in a greater number of ties, but not necessarily when the ties initiate at the node. The prestige cannot be quantified unless the network is directed. The difference between prestige and centrality is analogous to the difference between outdegree and indegree.

Eigenvector Centrality This indicates the importance of a node in the network. Relative scores are assigned to all nodes in the network based on the rule that ties to high-scoring vertices add more to the score of a node compared with the equal connections to low-scoring nodes. "Katz centrality" and Google's "PageRank" are variants of eigenvector centrality.

Geodesic Distance This can be defined as the length of the shortest path between two particular nodes and thus reflects the minimum number of connections required to connect them.

Reach Two nodes of a network are reachable if a path (and thus a walk) exists between them.

Clustering Coefficient This measures the likelihood that two neighbours of a node are also neighbours of each other. The higher the clustering coefficient, the greater the level of "cliquishness".

Cohesion This may be defined as the extent to which nodes are connected directly to each other through cohesive bonds. Groups are identified as "cliques" if every node is directly connected to every other node, "social circles" if there is less rigidness of direct interaction, or if precision is wanted, then groups are "structurally cohesive blocks".

3 Community Detection in Social Networks

Community detection (CD) in a social network has emerged as a fundamental tool for social network analysis that lays the foundation for several higher level analytics. Although nowadays the automation of community detection has become a favourite topic of research in the field of computational intelligence, it has already been a longstanding topic of research in sociology, as it has been associated with the study of urbanization, social marketing, criminology and various other areas [8]. It has many applications in biology and in physics. The term "community" first appeared in the book entitled "Gemeinschaft und Gesellschaft" published in 1887 [10].

Moreover, with the widespread use of online social networks such as Facebook, the study of the formation and evolution of community structures has gained greater significance. It is important to study community detection as it may prove to have many applications. For example, it may be used for targeted marketing for improved profit or influential targeted campaigning by election candidates. The study of community identification is based on the network structure [8]. It is associated with partitioning a social network into various cohesive subgroups or clusters in such a way that there is a strong cohesion or interconnection among the nodes of a subgroup, as represented by solid lines between nodes in Fig. 1 and weak cohesion among the nodes of different subgroups, represented by dashed lines in the figure. Figure 1 demonstrates three communities existing in the network [22]. The nodes in a community may have dense interaction owing to inter-personal relationships such as friendship and family, or business relationships etc.

As per Newman [26], the task of dividing a network into various communities by partitioning the nodes of a network into "g" number of groups, whereas minimizing the number of intra-group edges is referred to as "graph partitioning".

Many approaches have been used for community detection in social networks. These include the methodologies and the tools from various other disciplines such as physics, biology, sociology, computer science and applied mathematics. However, none of these algorithms solves the purpose and is not fully reliable. Fortunato [15] performed a comprehensive survey of "community detection in graphs" in 2010. The latest advanced review on community detection was carried out by Poonam Bedi and Chhavi Sharma in 2016. Other reviews available in the literature are

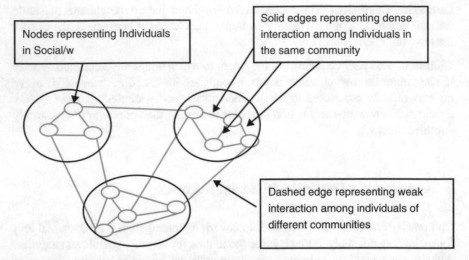

Fig. 1 A visual demonstration of community structure in a network

Lancichinetti and Fortunato [20] in 2010, Danon et al. [31] in 2005, and Khatoon and Aisha Banu [18] in 2015.

3.1 Algorithms for Community Detection

We can broadly categorize the community detection algorithms as follows:

1. **Graph partitioning algorithms:** These are associated with creating some predefined number of partitions of a graph with specific characteristics [32]. A common characteristic to be minimized is the "cut size". A cut is the splitting of the set of the nodes of a graph into two separate subsets and the number of edges between the subset gives the size of the cut. The Kernighan-Lin algorithm is one of the earliest heuristic algorithms for graph partitioning with time complexity as $O(n^3)$.

   This algorithm was designed to assign the "electronic circuits components to circuit boards to reduce the number of connections between boards" [5, 32]. The algorithm is aimed at maximizing Q, the difference between the number of intra-component edges and the number of inter-component edges. Initially, the algorithm partitions the graph into two components of a predefined size, either at random or by using a specific property of the graph. Then, pairs of vertices are swapped until a maximum value of Q is reached. Some of these swaps are done to escape local maxima.

2. **Hierarchical clustering algorithms:** Social networks often exhibit an interesting property of being a hierarchical structure with smaller communities lying under larger communities. This property calls for hierarchical community

detection algorithms that can unveil the hierarchy of communities in a social network. These algorithms are further categorized as:

(a) **Agglomerative hierarchical algorithms**
(b) **Divisive algorithms**

Agglomerative hierarchical algorithms, $O(n^3)$, follow the bottom up approach in which two lower level communities combine or merge to form a higher-level community if they are similar to each other [34]. These algorithms begin by considering each vertex as a community of its own and proceed upward until these are stopped by some stopping criteria, failing which these algorithms end up with the whole graph as a community Fig. 2. On the contrary, Divisive algorithms, as represented in Fig. 3, follow the top-down approach and split communities iteratively into two parts by removing edges between the least similar vertices. It starts with the whole graph as one community and successively

Agglomerative Algorithm 1: A Generic Agglomerative Algorithm

input : An input graph G

output: A partition into communities

Compute similarity of every vertex pair;

Put every pair of vertices with corresponding similarity value into a priority queue Q;

Start with every vertex as a community on its own;

while stopping condition not met do

    if Q is empty then

        break;

    if current partition satisfies the specified requirements then

        break;

Remove the most similar pair of vertices from Q and merge their corresponding communities into a single community;

    Return the computed partition into communities:

Fig. 2 Pseudo code for the hierarchical agglomerative clustering algorithms

Algorithm 2: A generic divisive algorithm

input : An input graph G

output: A partition into communities

//Compute similarity of every vertex pair

//Put every pair of vertices with corresponding similarity value into

a priority queue Q//

1. Start with all vertices as a single community
2. while stopping condition not yet met do

   if Q is empty then

   break

   if current partition satisfies the specified requirements

   then

   break
3. Remove the least similar pair of vertices from Q and remove

   the edge between them

Return the computed partition into communities

Fig. 3 Pseudo-code for hierarchical divisive algorithm

divides the node set into two subsets until the stopping criterion is met. The most common divisive algorithm is the Girvan–Newman (GN) algorithm [26]. This algorithm depends on the edge-betweenness, i.e. finding the edges that are the most obvious choices for being "between" communities or in other words, finding "inter-community" edges. The edge-betweenness algorithm for sparse graphs takes $O(m^2 n)$ and is $O(n^3)$ time, where m is the number of edges, and n denotes the number of nodes. The edge-betweenness algorithm runs faster than random walk betweenness or current-flow [$O(n^2)$ versus $O(n^3)$] on sparse graphs. In addition, the Girvan–Newman algorithm with edge betweenness gives better results in practical applications than those obtained by adopting other centrality measures [41].

Although hierarchical clustering-based algorithms have proved to be better than the graph partition-based clustering algorithms, these algorithms do not perform global optimization of the objective function. Instead, hierarchical clustering techniques perform local optimization of the objective function and decide which of the closest clusters are to be joined (or split in the case of divisive hierarchical clustering) at each step, where "closeness" is specified by specific measures of proximity. The results of these algorithms also depend on the conceptual visualization of clusters and the cluster proximity is taken into account. Moreover, agglomerative hierarchical clustering techniques are adversely affected by the outliers and noise, non-convex shapes and are inclined to break large clusters.

3. **Greedy modularity-based algorithms:** Modularity has been the most widely used quality function for community detection. Several techniques based on modularity use the greedy approach, such as agglomerative hierarchical clustering based on modularity [26]. At the beginning, every vertex is considered to be a separate community, and communities are merged iteratively such that each merge is locally optimal (i.e. yields the largest increase in the current value of modularity). Modularity has been defined as

$$Q = \sum e_{ii} - a_i^2$$

where e_{ii} represents the fraction of the edges that connect nodes in community i, e_{ij} denotes the fraction of the edges that connect nodes in two different communities, i and j, and

$$a_i = \sum_i f_{ij}$$

is the fraction of edges that are incident on the vertices in community i. The value $Q = 1$ indicates a network with a strong community structure.

The algorithm stops when it has reached the maximum modularity, so that it gives a grouping and a dendrogram. However, it is known to suffer from a resolution limit, i.e. communities below a given size threshold (depending on the number of nodes and edges) are always merged with neighbouring communities. Other methods based on greedy techniques include simulated annealing, external optimization and spectral optimization.

4. **Community detection in weighted network:** Community identification in weighted networks uses the greedy approach, which emphasizes maximizing the net weight of all the selected clusters and minimizing the similarity among the selected clusters. Then the calculation of the total weight of the selected clusters and the similarity between them is done [29]. The limitations of this approach are that it assigns each node to exactly one cluster, thus ensuring that every cluster has at least one object. It also ascertains that particular numbers of clusters are

selected, and ensures that a cluster must be selected if a data object is assigned
to it [18].

5. **Clustering based on edge betweenness [9]:** This is a hierarchical decomposition methodology where edges are removed in the descending order of their edge-betweenness scores (the number of shortest paths having a particular edge). Although this algorithm gives good results, it is slow owing to the repeated calculation of the edge betweenness score after every edge removal [16]. The modularity may be represented as:

$$Q = \frac{1}{2m} \sum_{ij} \left[A_{ij} - \frac{(K_i K_j)}{2m} \right] \delta(c_i, c_j),$$

where $m = \frac{1}{2} \sum_{ij} A_{ij}$ represents the number of edges in the graph, while k_i, k_j are degrees of vertices i and j, and $\delta(u, v)$ is 1 if $u = v$ and 0 otherwise.

6. **Community detection based on random walk [9, 10]:** This method is based on "random walks on the graph". The basic idea is to perform a random walk on the graph that tends to stay within the same community because there are very few edges that can take us outside of the given community. The walktrap algorithm runs shorter walks of 3-4-5 steps depending on the parameter used and uses the result of these random walks to agglomeratively merge separate communities to form a larger community. It may be combined with the modularity score to select where to cut the dendrogram. It is a little slower than the fast greedy approach but is more accurate.

7. **Leading eigenvector-based community detection:** In 2006, Newman [27] introduced another approach to characterizing the modularity matrix in terms of eigenvectors and deducing modularity by using the following equation:

$$Q = \frac{1}{4m} s^T B s,$$

where the modularity matrix is given by $B_{ij} = \left[A_{ij} - \frac{K_i K_j}{2m} \right]$ and modularity is given by the eigenvectors of the modularity matrix. The algorithm runs in $O(n^2 \log n)$ time, where $\log n$ represents the average depth of the dendrogram.

8. **Label propagation-based community detection:** Raghavan et al. in 2007 proposed an algorithm to detect community structure in the network based on label propagation that runs in non-linear time and does not require any previous information about communities or optimization of the pre-defined objective function. In this algorithm, a label is provided to each node. It then proceeds iteratively and updates the labels to nodes in a way that each node is assigned the most frequently occurring label among its neighbours [31]. The process stops each node from possessing the most frequently occurring label in its neighbourhood. It is very fast but yields different results based on the initial configuration (which is decided randomly); therefore, the method should be

carried out a many times (say, 1000 times for a graph) and a consensus labelling should be built, which could be tedious.

3.2 Community Detection Algorithms for Overlapping Networks

Although, the field of community detection has witnessed a lot of work, unfortunately, most of the community detection algorithms assume that communities in a network are discrete, which is far from the case. In social networks, communities may be overlapping, i.e. a node may fall into more than one community. In other words, communities in real-life networks are not disjoint. For instance, a person on Facebook may belong to a family group, an old-friends group, a hobby group, a business group and many other groups. Thus, communities with this type of node could hardly be discrete.

On review, disjoint community detection was found to follow five research lines, i.e. modularity maximization, spectral clustering, differential equation, random walks, and statistical mechanics to identify communities. These classic algorithms for community detection may not work well for the real structure of social networks. Overlapping community detection may be categorized into two classes: clique percolation-based [37] and non-clique percolation-based [20, 21]. Palla et al. [28] provided a solution to an overlapping community detection problem based on a clique percolation [20, 21]. The algorithm first finds all cliques in the network and identifies communities using component analysis of a clique–clique overlap matrix [6]. The clique percolation method runs in $O(\exp(n))$ time. Non-clique-based methods include label propagation, random walk [6], link partitioning and line graph, fuzzy detection, local expansion and optimization, agent-based and dynamical algorithms [20, 21].

3.3 Community Detection Algorithms for Dynamic Networks

A network is static if its configuration does not change with time. A network that keeps changing its configuration is dynamic. Social networks are characterized by their dynamic nature as the membership of a node evolves and keeps changing over time. Thus, the dynamics of the network change over time. As per Fortunato [15] and Deepjyoti Choudhury [10], three algorithms have been provided for dynamic algorithm: the spin model, random walk and synchronization. Wolf et al. proposed a mathematical and computational model for the analysis of dynamic communities that was based on interactions among nodes in the network [6].

4 Nature Inspired Algorithms and Their Applications in Community Detection

Computational intelligence as shown in Fig. 4 refers to a set of evolutionary techniques and methods of addressing complicated real world problems for which conventional mathematical tools and techniques are ineffective. The methods used in computational intelligence are close to the human way of reasoning and uses combinations of the following main techniques:

Nature-inspired algorithm (NIA) has evolved as a promising solution for the optimization of hard problems in a reasonable manner, and thus has found a niche among researchers tackling hard real-world problems, i.e. the problems for which exact solutions do not exist, but for which an approximate solution may be found.

Nature-inspired algorithms are inspired by the various processes observed from nature. The motive of developing such computing methods is to optimize engineering problems such as numerical benchmark functions, multi-objective functions and solving NP-hard problems for higher dimensions and greater sizes of problem.

Nature-inspired computation may be categorized as in Fig. 5.

There has been many solutions to community detection in social networks that comes from the family of nature-inspired algorithms, i.e. genetic algorithm,

Fig. 4 Evolutionary techniques under computational intelligence

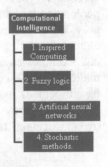

Fig. 5 Categories of nature-inspired computing algorithms

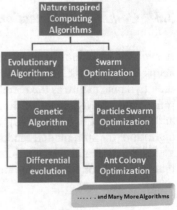

artificial bee colony optimization, ant lion optimization, hybrid of improved teacher learner based optimization (I-TLBO) and group search optimization (GSO) algorithms) [11] and multi-objective evolutionary algorithm etc. Thus, there is a good probability of nature-inspired algorithms tackling community detection problems in social networks efficiently.

The functioning of nature-inspired algorithms may be generalized in the following way to optimize any problem solution.

General Steps in Solving a Problem Using Nature-Inspired Algorithms [13]

1. Initialize population or search agents
2. Find the fitness function
3. Iterate until the stopping criteria are achieved or the maximum iterations are completed

   a. Calculate the fitness function for all search agents
   b. Find the best optimal fitness value
   c. Evaluate the new fitness value obtained against the previous one
   d. If the new fitness value is better than the previous value replace the fitness value with the newly obtained value
   e. Return the fitness value

4. Exit.

In this section we explore the application of various nature inspired algorithms in community detection.

Community detection methods have been proposed under two categories:

1. Community detection for unsigned networks
2. Community detection for signed networks

Communities of unsigned social networks can be identified as the group of nodes with more intense interconnections within the group and less intense interconnections among different communities. In recent years, a new type of social network has received a high level of attention from researchers across the globe, that is, the signed social network. In signed social networks, community structure is defined not only by the extent of interlinking but also by the signs of the links between two nodes. There are more positive signs between nodes in a community and more negative signs between nodes of two different communities. Positive edges represent the like, trust or affinity between two nodes or individuals, whereas negative edges represent the dislike, distrust or aversion between two nodes. Dataset for signed social networks may be found at Slashdot news review site, Wikipedia vote site and Epinions consumer review site [2].

4.1 Community Detection for Static, Unsigned and Non-overlapping Networks

Several evolutionary algorithms have been mapped to provide optimized solutions to community detection problems for static, unsigned and non-overlapping social networks. The following section provides a description of the applications of some important evolutionary algorithms.

4.1.1 A Nature-Inspired Metaheuristic Lion Optimization Algorithm

The lion optimization algorithm (LOA) was proposed by Maziar Yazdani and Fariborz Jolai in 2015. LOA is metaheuristic algorithm that falls under stochastic optimization and is characterized by the generation of different solutions for the problem in each run. The LOA is inspired by simulation of the special style exhibited by lions for capturing prey, territorial marking, migration, difference between the life style of nomad and resident lions. Lions thus have two types of social behaviour, nomadic and resident, and may switch the type of social organization. They live in groups called prides. The territory is the area in which a pride lives. The lions and the cubs protect the pride from nomadic lions. The LOA finds the optimal solution based on a lion's behaviour regarding territory defence and territory takeover. Through territory defence, the LOA evaluates the existing solution (territorial lion) and newly generated solution (nomadic lion) and the best one becomes the new solution, replacing the previous one. By territorial takeover, selection of the best territorial male is made from among the old territorial and new territorial males.

In 2015, Ramadan Babers et al. proposed a nature-inspired meta-heuristic LOA for community detection [2]. It used a normalized mutual index and modularity as the performance functions. Normalized mutual information (NMI), introduced by A. Lancichinetti, is a quality function and has become a standard metric for the evaluation of community detection algorithms. NMI is useful only when communities of vertices are known.

Given two partitions A and B of a network in communities, let C be the confusion matrix, whose element Cij is the number of nodes of the community i of partition A that are also in the community j of partition B. The normalized mutual information $I(A; B)$ is defined as:

$$\text{NMI}(A, B) = \frac{-2 \sum_{i=1}^{C_A} \sum_{j=1}^{C_B} c_{ij} \log \left(\frac{c_{ij}N}{c_{i.}c_{.j}} \right)}{\sum_{i=1}^{C_A} .C_{i.} \log \left(\frac{c_i}{N} \right) + \sum_{j=1}^{C_B} c_{.j} \log \left(\frac{c_j}{N} \right)}$$

where $C_A (C_B)$ is the number of groups in partition A (B), $C_{i.}(c_{.j})$ is the sum of the elements in row i (column j) and N is the number of nodes.

The experiments were performed on the Zachary Karate Club, the Bottlenose Dolphin Network, and the American College Football Network datasets. The results showed that the LOA seemed promising in terms of accuracy and quite successfully found optimized community structures for the presented datasets. The results obtained are summarized later in Table 1.

Table 1 Comparative analysis of evolutionary algorithm

(a) Modularity obtained by various algorithms

Reference	Methods	Modularity		
		Karate	Football	Dolphins
Proposed	GWO	0.4145	0.6038	0.5222
[16]	GN	0.38	0.577	0.495
[3]	TL-GSO	0.4185	0.6022	0.5019
[3]	GSO-1	0.3845	0.4276	0.4298
[3]	CNM	0.3806	0.5497	0.4954
[3]	Multilevel	0.4188	0.6045	0.5185
[11]	Discrete bat algorithm	0.4696	0.612	0.5498
[2, 38]	Lion optimization	0.46	0.89	0.59
[30]	MOGA-Net	0.4159	0.515	0.505
[1]	Bi-objective community detection	0.419	0.577	0.507
[3, 36]	Memetic algorithm	0.402	0.6044	0.5232

(b) NMI index obtained by various algorithms

Reference	Methods	NMI		
		Karate	Football	Dolphins
Proposed	GWO	0.8895	0.9734	0.8629
[16]	GN	0.692	0.762	0.573
[11]	Discrete bat algorithm	0.6873	0.84786	0.5867
[2, 38]	Lion optimization	0.67	0.91	0.58
[30]	MOGA-Net	1	0.796	1
[39]	MPCA	1	0.923	0.956
[30]	GA	0.818	0.805	0.935
[1]	Bi-objective community detect	0.695	0.878	0.615
[3, 36]	Memetic algorithm	0.693	0.762	0.573

(c) Number of communities identified by various algorithms

Reference	Methods	No. of communities		
		Karate	Football	Dolphins
Proposed	GWO	4	10	5
[3]	TL-GSO	4	10	5
[3]	GSO-1	3	7	5
[3]	CNM	3	6	4
[3]	Multilevel	4	10	5
[2, 38]	Lion optimization	4	10	5
[3, 36]	Memetic algorithm	3	7	4

4.1.2 TL-GSO: A Hybrid Approach to Mine Communities from Social Networks

The TL-GSO, proposed by Hema Banati and Nidhi Arora in 2015, is a hybrid algorithm based on the two latest evolutionary algorithms, the group search optimization (GSO) algorithm and the improved teacher learner-based optimization algorithm (I-TLBO). TL-GSO employs a single-point crossover and works on the same basis as GSO's Producer-Scrounger (PS) searching model, improving the earlier algorithm by employing multiple producers per generation instead of one, thus resulting in faster convergence of the best overall solution. TL-GSO was executed on four real-world data sets, i.e. the Karate Club Network, the Dolphin Network, the American Football Network and the Political Books Network. The results obtained from the algorithm have been summarized in Table 1.

4.1.3 Multi-Objective Genetic Algorithm

Pizzuti proposed a new algorithm for solve the problem by using a genetic algorithm. The algorithm uses the density measure and a latest concept of community score as a global measure to segment the network in clusters. The algorithm is aimed at maximizing this score [39]. The multi-objective genetic algorithm (MOGA-Net) is basically a non-dominated sorting genetic algorithm (NSGA-II) that builds and ranks the individual's population on a non-dominance basis [19]. The algorithm produces network communities at different levels of hierarchies, thus using a hierarchical clustering technique and automatically resolute communities are formed [24]. The author has used two competing objectives. The first one is about maximizing the interaction among the nodes in the same cluster and the second objective is to minimize the inert community interconnections. A multi-objective optimization problem $(\Omega, F_1, F_2, \ldots, F_t)$ is defined as

$$\min F_i(S), i = 1, \ldots, t \text{ subject to } S \epsilon \Omega$$

where $\Omega = \{S_1, S_2 \ldots, S_k\}$ is the set of plausible clustering of a network, and $F = \{F_1, F_2, \ldots, F_t\}$ is a set of t single criterion functions. Each $F_i : \Omega \rightarrow R$ is a distinct objective function for determining the feasibility of the clustering thus obtained.

The datasets used to test the algorithm are the Zachary Karate Club, the Bottlenose Dolphin Network, the American College Football Network and Krebs' Book. The pro side of the MOGA-Net algorithm is that it provides a set of solutions at different levels of the hierarchy by giving the opportunity to analyze the network structure at different resolution levels and hence proves to be better than the single objective approaches.

4.1.4 Multi-Population Cultural Algorithm

A cultural algorithm is an evolutionary framework based on knowledge and employs knowledge in the evolution of optimal solutions. Devised by Pooya Moradian Zadeh and Ziad Kobti in 2015 [39], the multi-population cultural algorithm (MPCA) is an algorithm in which population space representing the solution is randomly created based on the state space of the network. From among the initial solutions a belief space consisting of the solutions with better fitness values is generated. In each generation, the set of the best individuals of each population is used to update the belief space. The algorithm is evaluated against GA-Net, MOGA-Net, DECD (differential evolution) and the NMI value has been used as the performance evaluation function to measure the similarity between the newly detected community structure and the actual one. The experiment was performed on the synthetic network and on the real datasets of the Zachary Karate Club, Dolphin and American Football Networks. The results have been summarized in Table 1.

Other algorithms that have been applied for the community detection of unsigned, static and non-overlapping community structure in social networks include bi-objective community detection using the genetic algorithm proposed by Rohan Agrawal in 2011 [1], the GA-Net algorithm [30] proposed by Clara Pizzuti in 2008, the memetic algorithm for community detection in networks proposed by Maoguo Gong et al. in 2011 [17] and the multi-objective community detection based on memetic algorithm [36] proposed by Peng Wu and Li Pan in 2015.

4.2 Community Detection for Signed Networks

This section describes the applications of various evolutionary algorithms to signed social networks.

4.2.1 Cultural Algorithm Based on Artificial Bee Colony Optimization

The cultural algorithm based on artificial bee colony (ABC) optimization was proposed by HU Baofang in 2015. The experiment was conducted on synthetic signed networks generated with three different sizes [4] and two real-world networks (the Gahuku-Gama Sub tribes [GGS] network, the Slovene Parliamentary Party (SPP) network). The cultural algorithm is a knowledge- and belief-based evolutionary framework that employs knowledge to accelerate the evolution process and conduct the direction of evolution. Topological knowledge is employed in our proposed algorithm, which is very suitable for local searching for employees and on-looker bees. The objective function used for the experiment is extended modularity in signed networks. A single objective optimization problem $(\Omega; F)$ is used, as shown in the equation below.

$$\text{Min } F(S), \text{s.t } S \in \Omega$$

where $F(S)$ is an objective function to be optimized, and $\Omega = \{S1, S2, \ldots, Sm\}$ is the set of possible communities of a signed network. Fitness function was given by:

$$F(S) = \sum_{ij} \lambda A_{ij}^{-}\delta(c_i, c_j) + (1 - \lambda)A_{ij}^{+}(1 - \delta(c_i, c_j))$$

where A_{ij} represents the weight of the edge between vertices i and j. A_{ij}^{-} denotes the negative edges between i and j and A_{ij}^{+} is the positive edges between i and j. Here, c_i is the community number of nodes i and $\delta(c_i, c_j) = 1$, if $\sigma_i = \sigma_j$, else $\delta(c_i, c_j) = 0.\lambda$ is a parameter through which the two types of edges, positive and negative, can be balanced and $0 < \lambda \leq 1$. The performance evaluation of the algorithm is based on the NMI metric. The algorithm was compared with the ABC algorithm, GA-Net and MPCA for signed networks and the author showed that the cultural algorithm proved to be better than those algorithms.

4.2.2 Multi-Objective Evolutionary Algorithm Based on Similarity

The multi-objective evolutionary algorithm (MEA-SN) was proposed by Chenlong Liu, Jing Liu and Zhongzhou Jiang in 2014. The authors considered both the link density and the sign of the network. Two objectives are proposed. The first objective is to bring all positive links in communities, whereas the second is to keep all negative links as inter-community links. As the MEAs-SN can switch between different representations during the evolutionary process, it thus benefits from both representations and can detect both separate and overlapping communities [23]. The algorithm makes use of a modified modularity function for signed networks, represented by Q_{signed} and modified modularity function for overlapping community structures, represented by Q_{ov}. The authors generated synthetic networks using the Lancichinetti–Fortunato–Radicchi (LFR) benchmark, suitable for both separate and overlapping networks. The results are evaluated using NMI and modularity values, and are shown in Table 1.

4.2.3 Discrete Bat Algorithm

Inspired by the bat algorithm proposed by Xin-She Yang, Wang Chunyu proposed a novel discrete bat algorithm [11] in 2015. The authors of this algorithm adopted an ordered adjacency list as the encoded form, and used modular Q function as the objective function. The accuracy of division, i.e. the rate of correct partitioning, is used as the evaluation standard, and simulation is done by using MATLAB R2013a. The analysis of simulation results shows that the discrete bat algorithm is effective and efficient at detecting communities in complex networks. The results are presented in Table 1.

4.3 Community Detection for Overlapping Communities

4.3.1 Genetic Algorithm for Identifying Overlapping Communities in Social Networks

Brian Dickinson et al. proposed a genetic algorithm in 2013 for the identification of overlapping communities in social networks using an optimized search space [12]. The author presents two algorithms for community detection. One is the LabelRank algorithm, which, because of its deterministic nature, is restricted to very few candidate solutions. The other algorithm is the genetic algorithm, which employs a restricted edge-based clustering technique to identify overlapping communities (OGA) by maximizing a modified and efficient overlapping modularity function. The experiments were performed on several real-world datasets, including the Karate Network, the Pilgrim Network, Lesmis Network, Dolphin Network and Football Network. The results are compared against the speaker-listener label propagation algorithm (SLPA). The algorithms are analyzed using the modified Q function and the F-score.

4.3.2 A Dynamic Algorithm for Local Community Detection in Graphs

A dynamic algorithm for local community detection in graphs was proposed by Anita Zakrzewska and David A. Bader in 2015 [40]. It was based on the dynamic seed set expansion, which helps to incrementally update the community when the associated graph changes. The algorithm has severe shortcomings that the community may split apart, but the algorithm is not able to identify this as the community score remains unchanged. The performance improvement is greatest in cases of low latency updates. The dynamic method is faster than re-computation, the experiment was conducted on the Slashdot graph, and precision and recall measures are used for performance evaluation. All these algorithms have been summarized in tabular form in the Appendix.

5 A Novel Approach to Community Detection Using Grey Wolf Optimization

In this section, we first briefly describe a novel algorithm for optimization, i.e. grey wolf optimization (GWO) proposed by Syed Mirzalili et al. in 2014 [25]. This is followed by a novel solution to community detection that uses GWO as a tool for optimization.

5.1 Grey wolf optimization

The GWO algorithm is inspired by the leadership hierarchy and hunting mechanism of grey wolves found in the mountains, forests and plains of North America, Europe and Asia. Grey wolves are characterized by a bushy tail and powerful teeth. They live and hunt in groups called packs. The average group size is 5–12. They exhibit an attractive social organization, with four categories of wolves, i.e. alpha, beta, delta and omega, and there is a leadership hierarchy in the pack.

(a) α is the dominant leader and decision maker in the pack.
(b) β & δ—the betas are subordinate to the alpha wolf and help the alpha in decision-making. The beta wolf is probably the next best candidate to be the alpha and can be a male or a female. Delta wolves must follow the alpha and betas, but they dominate the omegas.
(c) Ω-Omegas are the lowest ranking wolves in the pack. They are the scapegoats in the pack and always have to submit to the alphas, betas and deltas.

Apart from exhibiting the social hierarchy, they also exhibit other important social behaviour of particular interest, such as group hunting. The grey wolves perform group hunting, which may be divided into the following phases [25]:

1. Tracking, chasing and approaching the prey
2. Pursuing, encircling and harassing the prey until it stops moving
3. Attacking the prey.

The mathematical model of the encircling behaviour is represented by the equations:
equations

$$\mathbf{D} = |\mathbf{C}\mathbf{X}_p - \mathbf{A}\mathbf{X}(t)| \tag{1}$$

$$\mathbf{X}(t+1) = \mathbf{X}_p(t) - AD \tag{2}$$

where X_p is the position vector of the prey, X indicates the position vector of a grey wolf and A and C are coefficient vectors given by

$$\mathbf{A} = 2\mathbf{a}\mathbf{r}_1 - \mathbf{a} \tag{3}$$

$$\mathbf{C} = 2\mathbf{r}_2 \tag{4}$$

where,
   T is the current iteration
   $\mathbf{X}$ is the position vector of a wolf, and
   r1 and r2 are random vectors ϵ [0,1] and **a** vary linearly from 2 to 0.

5.1.1 The Mathematical Model for Hunting

The hunting mechanism of grey wolves may be modelled by the following equations:

$$\mathbf{D}_\alpha = |\mathbf{C}_1 \cdot \mathbf{X}_\alpha(\mathbf{t}) - \mathbf{X}(\mathbf{t})|, \mathbf{D}_\beta = |\mathbf{C}_2 \cdot \mathbf{X}_\beta(\mathbf{t}) - \mathbf{X}(\mathbf{t})|, \mathbf{D}_\delta = |\mathbf{C}_3 \cdot \mathbf{X}_\delta(\mathbf{t}) - \mathbf{X}(\mathbf{t})| \tag{5}$$

$$\mathbf{X}_1 = \mathbf{X}_\alpha(\mathbf{t}) - \mathbf{A}_1.(\mathbf{D}_\alpha), \mathbf{X}_2 = \mathbf{X}_\beta(\mathbf{t}) - \mathbf{A}_2.(\mathbf{D}\beta), \mathbf{X}_3 = \mathbf{X}_\delta(\mathbf{t}) - \mathbf{A}_3.(\mathbf{D}_\delta) \tag{6}$$

$$\mathbf{X(t+1)} = \frac{(\mathbf{X}_1 + \mathbf{X}_2 + \mathbf{X}_3)}{3} \tag{7}$$

where t indicates the current iteration, $X_\alpha(t)$, $X_\beta(t)$ and $X_\delta(t)$ are the position of the grey wolf α, β and δ at tth iteration, $X(t+1)$ presents the position of the grey wolf at $(t+1)$th iteration.

$$\mathbf{A(.)} = 2 \cdot a \cdot \text{rand}(0, 1) - a \tag{8}$$

$$\mathbf{C(.)} = 2 \cdot \text{rand}(0, 1) \tag{9}$$

Where "a" is the linear value that varies from 2 to 0 according to iteration. A(.) and C(.) are the coefficient vectors of the α, β and δ wolves. The GWO algorithm allows its search agents to update their position based on the location of α, β and δ and attack the prey. Although the proposed encircling mechanism demonstrates exploration to some extent, it is prone to stagnation in local solutions with these operators and hence needs more operators for better exploration.

5.2 Community Detection Using Grey Wolf Optimization

The pseudo code for the implementation of the GWO algorithm for community detection is presented below.

Pseudo code for the GWO algorithm

1. Initialize the alpha, beta and delta positions
2. Initialize the position of the search agents
3. Loop until maximum iteration is not reached:

   (a) Calculate the objective function for each search agent
   (b) Update alpha, beta, delta and omega positions

(continued)

(c) Evaluate the new fitness value obtained against the previous one
(d) Return the search agents that go beyond the boundaries of the search space.

4. Return the solutions.

We have used the modularity function as the objective function and have used NMI as the performance evaluation criteria together with the modularity of the network.

5.3 Experiment

The GWO algorithm was used for community detection for the first time. The experiment was carried out on Intel(®) core(^{TM}), i5 CPU @ 2.53 GHz, 4 GB RAM using MATLAB R2010a. The GWO algorithm was used to optimize the modularity function, and uses NMI index values and modularity values as the performance evaluation criteria.

Datasets We used three real-world datasets for the purposes of community detection.

1. The Zachary Karate Club Network, consisting of 34 members represented by the vertices and the relationships among the club president and the karate instructor represented by edges. It simulates the relationships between the club president and the karate instructor and contains 78 edges [2].
2. The Bottlenose Dolphin Network, simulating the behaviour of bottlenose dolphins seen over 7 years. It contains 62 nodes and 318 edges.
3. American College Football Network [2] demonstrates the football games between American Colleges during a regular season in fall 2000. It contains 115 nodes and 1226 edges.

Parameter Initialization

```
No. of search agents : 50
Maximum no. of iteration : 50
No. of trial runs : 10
Lower bound =0, upper bound = 10 (for the Karate Network)
Lower bound =0, upper bound = 20 (for the Football Network)
Lower bound =0, upper bound = 10 (for the Dolphin Network)
```

Experimental Results The GWO algorithm was first proposed for community detection and was implemented on three different real-world datasets for the extraction of community structure in those real networks. The GWO algorithm was executed for ten trial runs for each dataset. The modularity and NMI values obtained across the ten trial runs were averaged to provide the final modularity and NMI values, as shown in Table 1a and b respectively.

The results of modularity value and NMI values thus obtained from the experimental setup are further compared with the state of the art algorithms that were employed earlier for community detection. These include TL-GSO, the discrete bat algorithm, lion optimization, MOGA-Net etc.

For better analysis, we compare our results of average modularity and average NMI with those algorithms that have also employed similar criteria. The source of modularity values obtained from various algorithms is mentioned in the first column. The results are summarized under Table 1a for modularity value, Table 1b for comparative analysis of NMI values and Table 1c for the number of communities detected by various algorithms.

Figure 6 shows that the GWO algorithm demonstrates competitive modularity compared with the GN, TL-GSO, lion optimization and memetic algorithms, bi-objective community detection (BOCD) and others, with discrete bat algorithm and lion optimization algorithms as the exceptions.

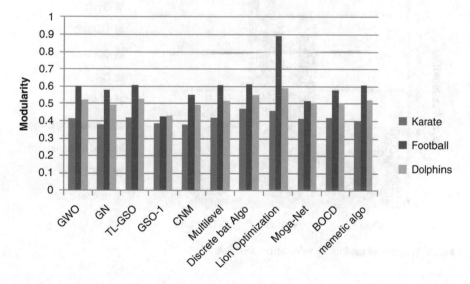

Fig. 6 Modularity comparison

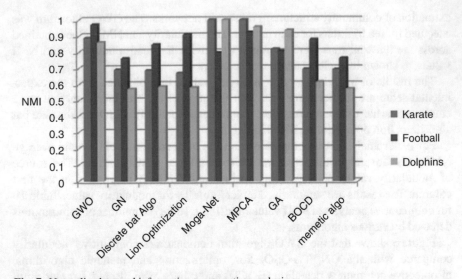

Fig. 7 Normalized mutual information comparative analysis

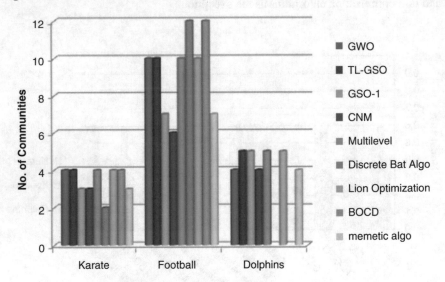

Fig. 8 Variation of modularity value across ten trials

Figure 7 demonstrates the comparison of NMI values obtained through the GWO algorithm with lion optimization, discrete bat algorithm and TL-GSO algorithms, MOGA-Net and other contemporary evolutionary algorithms.

Figure 8 demonstrates the number of communities identified by various evolutionary algorithms for the three different datasets, compared with our proposed algorithm.

Table 2 Experimental results for the GWO algorithm

Dataset	Modularity			NMI		
	Highest	Lowest	Average	Highest	Lowest	Average
Karate	0.4198	0.3942	0.4145	1	0.815	0.908
Football	0.6046	0.5976	0.6038	1	0.9199	0.9734
Dolphin	0.5277	0.5188	0.5222	1	0.7224	0.863

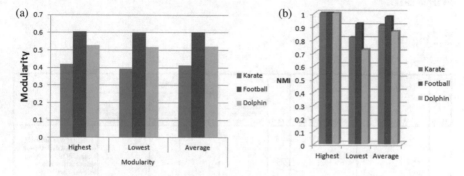

Fig. 9 Experimental results for community detection using the GWO algorithm. (**a**) Modularity results for GWO. (**b**) NMI results for the GWO algorithm

The lowest, highest and average values of modularity and NMI obtained through ten runs of the GWO algorithm for each dataset have been demonstrated in Table 2 and can be modelled by Fig. 9a and b respectively.

6 Conclusion

The experiments were performed on the three different datasets, i.e. the Karate Club Network (with 34 nodes and 78 edges), the Bottlenose Dolphin Network (with 62 nodes and 159 edges) and the American College Football Network(with 115 nodes and 613 edges) using the GWO algorithm for 10 runs. The highest, lowest and average values for the modularity and the NMI thus obtained have been summarized in Table 2. The highest modularity value has been reported to be 0.4198 for the Karate Club dataset, 0.6046 for the Football dataset and 0.5277 for the Dolphin dataset respectively, which is on a par with other evolutionary algorithms. The GWO algorithm resulted in four communities for the Karate Club dataset, ten communities for the Football dataset and five communities for the Dolphin dataset.

The results demonstrate the promising nature of the GWO algorithm. The algorithm does not need the number of communities of the dataset in advance. It optimizes the objective function at each iteration to finally converge at the final result. The algorithm has a higher rate of convergence. It converged in less than 50 iterations compared with other nature-inspired algorithms that took 100 s of iterations to converge.

Hence, the GWO algorithm is capable of generating the optimized results in a shorter period. The algorithm results in higher values of modularity than other evolutionary algorithms. The NMI values obtained from the GWO algorithm almost approach 1, which shows the accuracy of the results.

Appendix

See Fig. 10.

S.No	Algorithm	Author / Authors	Year	Performance evaluation parameters	Community Detection			Data sets
					Signed / Unsigned	Static / Dynamic	Overlapping/ Non-overlapping	
1	ABC based cultural algorithm	HU Baofang	2015	NMI	Signed	Static	Non-overlapping	artificial signed networks generated with 3 different sizes, and two real world networks (Slovene Parliamentary Party network (Spp), the Gahuku-Gama Subtribes netw3.ork (GGS)).
2	LION optimization algorithm	Ramadan Babera*, Aboul Ella, Hassanien, and Neveen I. Ghali	2015	NMI, community fitness, modularity	Unsigned	Static	Non-overlapping	Zachary kate Club, Bottlenose Dolphin network, American college Football Network
3	MEA-SN	Chenlong Liu, Jing Liu, Zhongzhou Jiang	2014	NMI, modularity	both signed and unsigned	Static	both overlapping and non-Overlapping	Synthetic networks are generated of size 1000, 5000, and 10000 nodes using (LFR) benchmark Real network- 1. Slovene parliamentary party Network, 2. Gahuku-Gama Subtribes network, 3. Karate Club network 4. Dolphin network, 5. Football Network
4	Discrete Bat Algorithm,	Wang chuyu* and Pan Yun	2015	accuracy of division, Modularity	unsigned	Static		Zachary Kate Club
5	GWO	Seyedali Mirjalili, Seyed Mohammad Mirjalili, Andrew Lewis	2014	NMI, Modularity function value	Unsigned	Static	Non Overlapping	Zachary Kate Club, Bottlenose Dolphin network, American college Football Network
6	TL-GSO	Hema Banati, Nidhi Arora	2015	NMI, Modularity	Unsigned	Static	Non Overlapping	Karate Club, Dolphin, Football, Political Books
7	Multiobjective Based Genetic Algorithm (MOGA-Net)	Clara Pizzuti	2009	Modularity and NMI	Unsigned	Static	Non Overlapping	Karate Club, Dolphin, Football, Krebs' Book
8	Multi Population Cultural Algorithm (MPCA)	Pooya Moradian Zadeh, Ziad Kobti	2015	NMI	Unsigned	Static	Non Overlapping	Karate Club, Dolphin, Football
9	A Genetic Algorithm for Identifying Overlapping communities (OGA)	Brian Dickinson, Benjamin Valyou, Wei Hu	2013	Modified modularity (Q), F-Score	Unsigned	Static	Overlapping	Karate Network, Pilgrim Network, Lermis, Dolphin and Football Network.
10	A Dynamic Algorithm for Local Community detection in graphs	Anita Zakrzewaka and David A. Bader	2015	Precision and recall	Unsigned	Static	Overlapping	slashdot graph

Fig. 10 Summary of evolutionary algorithms in Community detection

References

1. Agrawal, R.: Bi-objective community detection (BOCD) in networks using genetic algorithm. In: Communications in Computer and Information Science, vol. 168, September 2011. doi:10.1007/978-3-642-22606-9_5. Source: arXiv
2. Babers, R., et al.: A nature-inspired metaheuristic lion optimization algorithm for community detection. In: Computer Engineering Conference (ICENCO), 2015 11th International, IEEE (2015). 978-1-5090-0275-7/15/2015
3. Banati, H., Arora, N.: TL-GSO - a hybrid approach to mine communities from social network. In: 2015 IEEE International Conference on Research in Computational Intelligence and Communication Networks (ICRCICN)
4. Baofang, H.U.: A cultural algorithm based on artificial bee colony optimization for community detection in signed social networks. In: 2015 10th International Conference on Broadband and Wireless Computing, Communication and Applications (2015)
5. Barbara, D.: An Introduction to Cluster Analysis for Data Mining (2000). Retrieved November 12, 2003, from http://www-users.cs.umn.edu/~han/dmclass/cluster_survey_10_02_00.pdf
6. Bedi, P., Sharma, C.: Community detection in social networks. Adv. Rev. Published online: 19 Feb 2016. doi:10.1002/widm.1178. WIREs Data Mining Knowledge Discovery. doi:10.1002/widm.1178
7. Bhawsar, Y., Thakur, G.S.: Community detection in social networking. J. Inf. Eng. Appl. 3(6) (2013). www.iiste.org. ISSN: 2224-5782 (print). ISSN: 2225-0506 (online) - Selected from International Conference on Recent Trends in Applied Sciences with Engineering Applications
8. Chen, W.: Community detection in social networks through community formation games. In: Proceedings of the Twenty-Second International Joint Conference on Artificial Intelligence (2011)
9. Chen, F., Li, K.: Detecting hierarchical structure of community members in social networks. Knowl.-Based Syst. 87, 3–15 (2015)
10. Choudhury, D., Paul, A.: Community detection in social networks: an overview. Int. J. Res. Eng. Technol. eISSN: 2319-1163. pISSN: 2321-7308
11. Chunyu, W., Yun, P.: Discrete bat algorithm and application in community detection. Open Cybernet. Syst. J. 9, 967–972 (2015)
12. Dickinson, B.: A genetic algorithm for identifying overlapping communities in social networks using an optimized search space. Soc. Netw. 2, 193–201 (2013). http://dx.doi.org/10.4236/sn.2013.24019. Published Online October 2013 http://www.scirp.org/journal/sn
13. Dixit, M., et al.: An exhaustive survey on nature inspired optimization algorithms. Int. J. Softw. Eng. Appl. 9(4), 91–104 (2015). http://dx.doi.org/10.14257/ijseia.2015.9.4.11
14. Eindride, E.: Community Detection in Social Networks, a thesis. Department of Informatics, University of Bergen (2015)
15. Fortunato, S.: Community detection in graphs. arXiv:0906.0612v2 [physics.soc-ph] 25 Jan 2010
16. Girvan, M., Newman, M.E.J.: Community structure in social and biological networks. PNAS 99(12), 7821–7826 (2002)
17. Gong, M.: Memetic algorithm for community detection in networks. Phys. Rev. E 84, 056101 (2011)
18. Khatoon, M., Aisha Banu, W.: A survey on community detection methods in social networks. Int. J. Educ. Manage. Eng. 1, 8–18 (2015)
19. Kim, K., et al.: Multiobjective evolutionary algorithms for dynamic social network clustering. In: GECCO 10 Proceedings of the 12th Annual Conference on Genetic and Evolutionary Computation, pp. 1179–1186
20. Lancichinetti, A., Fortunato, S.: Community detection algorithms: a comparative analysis. arXiv:0908.1062v2[physics.soc-ph]16 Sep 2010
21. Lancichinetti, A., Radicchi, F., Ramasco, J.J., Fortunato, S.: Finding statistically significant communities in networks. PLoS One 6(4), e18961 (2011). doi:10.1371/journal.pone.0018961

22. Li, K., Pang, Y.: A unified community detection algorithm in complex network. Neurocomputing **130**, 36–43 (2014). www.elsevier.com/locate/neucom
23. Liu, C.: A Multiobjective Evolutionary algorithm based on similarity for community detection from signed social networks. IEEE Trans. Cybernet. **44**(12) (2014)
24. Mahajan, A., Kaur, M.: Various approaches of community detection in complex networks: a glance. Int. J. Inf. Technol. Comput. Sci. **4**, 35–41 (2016). Published Online April 2016 in MECS. http://www.mecs-press.org/DOI:10.5815/ijitcs.2016.04.05
25. Mirjalili, S.: Grey wolf optimizer. Adv. Eng. Softw. **69**, 46–61 (2014)
26. Newman, M.E.J.: Detecting community structure in networks. Eur. Phys. J. B **38**, 321 (2004). doi:10.1140/epjb/e2004-00124-y
27. Newman, M.E.J.: Finding community structure in networks using the eigenvectors of matrices. Phys. Rev. E **74**, 036104 (2006)
28. Palla, G., Derenyi, I., Farkas, I., Vicsek, T.: Uncovering the overlapping community structure of complex networks in nature and society. Nature **435**(7043), 814–818 (2005)
29. Pizzuti, C.: A multi-objective genetic algorithm for community detection in networks. In: ICTAI 2009, 21st IEEE International Conference on Tools with Artificial Intelligence, Newark, NJ, 2–4 November 2009
30. Pizzuti, C.: A multiobjective genetic algorithm to find communities in complex networks. IEEE Trans. Evol. Comput. **6**(3), 418–430 (2012)
31. Raghavan, U.N., et al.: Near linear time algorithm to detect community structures in large-scale networks. Phys. Rev. E **76**, 036106 (2007)
32. Social Network Analysis: Theory and Applications. https://www.politaktiv.org/documents/10157/29141/SocNet_TheoryApp.pdf
33. Steinhaeuser, K., Chawla, N.V.: Community detection in a large real-world social network. In: Social Computing, Behavioral Modeling, and Prediction, pp. 168–175
34. Wang, M., et al.: Community detection in social networks: an indepth benchmarking study with a procedure oriented framework. In: Proceedings of the VLDB Endowment VLDB Endowment Homepage archive, vol. 8(10), June 2015, pp. 998–1009
35. Wasserman, S., Faust, K.: Social Network Analysis. Cambridge University Press, Cambridge (1994)
36. Wu, P., Pan, L.: Multi-objective community detection based on memetic algorithm. PLoS One **10**(5), e0126845. doi:10.1371/journal. pone.0126845 (2015)
37. Xu, Y.: Finding overlapping community from social networks based on community forest model. Knowl.-Based Syst. 1–18 (2016). ISSN: 0950-7051 000
38. Yazdani, M., Jolai, F.: Lion optimization algorithm (LOA): a nature-inspired metaheuristic algorithm. J. Comput. Des. Eng. **3**, 24–36 (2016)
39. Zadeh, P.M., Kobti, Z.: A multi-population cultural algorithm for community detection in social networks. In: The 6th International Conference on Ambient Systems, Networks and Technologies (ANT 2015), Procedia Computer Science, vol. 52, pp. 342–349 (2015)
40. Zakrzewska, A., Bader, D.A.: A dynamic algorithm for local community detection in graphs. In: 2015 Proceedings of the 2015 IEEE/ACM International Conference on Advances in Social Networks Analysis and Mining 2015, ASONAM 2015, Paris, 25–28 August 2015, pp. 559–564
41. Newman, M.E., Girvan, M.: Finding and evaluating community structure in networks. Phys. Rev. E **9**(2), 026113 (2004)

A Unified Framework for Community Structure Analysis in Dynamic Social Networks

Sajid Yousuf Bhat and Muhammad Abulaish

Abstract One of the major tasks related to structural social network analysis is the detection and analysis of community structures, which is highly challenging owing to consideration of various constraints while defining a community. For example, community structures to be detected may be disjointed, overlapping, or hierarchical, whereas on the other hand, community detection methods may vary depending on whether links are weighted or unweighted, directed or undirected, and whether the network is static or dynamic. Although a number of community detection methods exist in literature, most of them address only a particular aspect, and generally, community structures and their evolution analysis are studied separately. However, analyzing community structures and their evolution under a unified framework could be more useful, where community structures provide evidence of community evolution. Moreover, not many researchers have dealt with the issue of the utilization of detected communities and they have simply proposed methods to detect communities without emphasizing their utilities. This chapter presents a unified social network analysis framework that is mainly aimed at addressing the problem of community analysis, including overlapping community detection, community evolution tracking, and hierarchical community structure identification in a unified manner. At the end of this chapter, we present some important application areas in which the knowledge of community structures facilitates the generation of important analytical results and processes to help solve other social network analysis-related problems. The application areas mainly include dealing with spammers, in which we present the importance of various community-based features in learning predictive models for spammer detection in online social networks. In addition, we address the issue of detecting deceptions in online social networks, which mainly includes dealing with cloning attacks, and the importance of community detection for facilitating the process of viral marketing in online social networks.

S.Y. Bhat
Department of Computer Sciences, University of Kashmir, Srinagar, India
e-mail: s.yousuf.jmi@gmail.com

M. Abulaish (✉)
Department of Computer Science, South Asian University, New Delhi, India
e-mail: abulaish@ieee.org

© Springer International Publishing AG 2017
H. Banati et al. (eds.), *Hybrid Intelligence for Social Networks*,
DOI 10.1007/978-3-319-65139-2_4

Keywords Community analysis applications • Community detection • Community evolution tracking • Hierarchical and overlapping community • Social network analysis

1 Introduction

Social networks resemble graph structures and are aimed at modelling relationships/ties reflecting associations, friendship, hyperlinks, financial exchange, co-authorship, citations, interactions etc., between social actors such as people, web, pages, research articles, financial accounts, airports, etc. [28]. Social networks have been an interest of study since the mid-1930s starting with the introduction of *Sociometry* by Moreno [22], and later in 1950s, the applications of graph theory started becoming popular in sociological community [27]. Some of the most important characteristics of social networks that motivated for significant interest and developments in the field of social network analysis are listed below:

Power law and preferential attachment: Researchers such as Barabási and Albert [2] revealed that unlike random networks, the degree distribution of nodes in a large-scale real-world network follows a scale-free power law, making some nodes have a relatively higher degree (forming hubs) with most nodes having fewer connections. Moreover, the probability of newly joined nodes to attach (form links) to nodes of a higher degree is greater than to nodes of a lower degree, thus making the growth of such networks follow a rich-get-richer scheme (aka preferential attachment) [3].

Assortativity and network clustering: Newman and Park [23] analyzed that identical nodes in terms of degrees in a social network tend to show more connectivity with each other than with others, thus highlighting positive correlations among the degrees of adjacent vertices (assortativity). Moreover, social networks have non-trivial clustering of nodes or network transitivity, making them exhibit communities, i.e. nodes within a group have a high similarity or connectivity than the remaining nodes of the network.

Small-world behavior and the strength of weak ties: Milgram [21] popularized the concept of the *six-degrees-of-separation* based on the analysis that the average path length between any two individuals in a society world is six hops, thus reflecting the small-world behavior of the social networks. The *weak ties* hypothesis proposed by Granovetter [10] is another famous concept of sociology, which reflects that the degree of neighborhood overlap between two individuals increases in proportion to the strength of their tie. This reveals that strong ties can play an important role in forming tight clusters, whereas weak ties play an important role in the flow of information and innovation in the network, which is often termed the *strength of weak ties*.

Based on the characteristics mentioned above, social network analysis (SNA) has its roots in various disciplines, including sociology, anthropology, economics, biology, communication, and social computing. The process of SNA mainly focuses on understanding the behavior of individuals and their ties in social networks that translate them into large-scale social systems. Because of the advent of Web2.0 and the existence of many online social networks (OSNs), the application of SNA has recently gained much popularity owing to its ability to model and analyze various processes taking place in society, such as recommendations, spread of cultural fads or diseases, community formation, etc. Moreover, the existence of huge amounts of both structural and non-structural data (aka user-generated content) provides an opportunity to analyze them for various purposes, including open-source intelligence, target marketing, business intelligence, etc. Besides traditional OSNs, a number of complex networks, such as co-authorship networks, citation networks, protein interaction networks, metabolic pathways, and gene regulatory networks also possess similar patterns that can be analyzed using SNA techniques. Keeping in mind the huge amount of data associated with social networks, information extraction, data mining, and NLP techniques have emerged as key techniques for analyzing social networks at different levels of granularity. As a result, a significant increase has been seen in the research literature at the intersection of computational techniques and social sciences. The exponential growth of OSNs and the large-scale complex networks induced by them provide unique challenges and opportunities in the area of social network analysis that can be broadly classified into *structural*, *non-structural*, and *hybrid* categories, depending on the nature of the data being used for analysis. Figure 1 presents a categorization of SNA tasks that are briefly described in the following sub-sections.

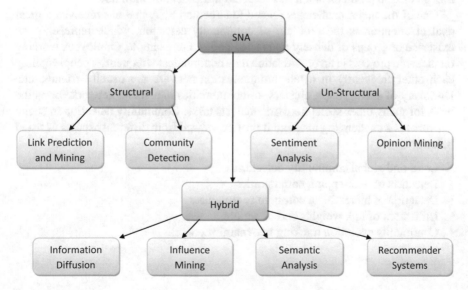

Fig. 1 Classification of social network analysis tasks

1.1 Structural Social Network Analysis

Structural social network analysis is mainly aimed at analyzing and mining structural (aka topological) data, such as the link structure of the WWW, social network users' interactions and friendship networks, citation networks, etc. Numerous graph mining techniques have been used for the analysis of such networks, which reveal important patterns and characteristics of human social behavior that are often highly surprising. One of the most primitive and challenging tasks in this domain is *link mining*. Introduced by Getoor [9], link mining is a data mining task that mainly deals with data related to social networks, which tend to be heterogeneous, multi-relational, or semi-structured. Link mining is an emerging area that explicitly models links between data instances to find non-trivial interesting patterns in the data. This field of research encompasses a range of tasks, including descriptive and predictive modeling, e.g., estimating the number of links in a network, predicting the relation type between two entities, estimating the likelihood of a link's existence, etc. Motivated by the dynamics and evolving nature of social networks, a sub-challenge sprouting from the domain of link mining is the problem of *link prediction*, which is mainly aimed at predicting possible future links, given the present state of the nodes and links in a social network. The heterogeneity of real-world social networks, which involves different node types, link attributes, weights and directions, friend and foe relations, etc., poses a major challenge to link prediction. Another great challenge is the prediction of links that could be induced in a social network when new nodes are added. This is referred to as the cold start link prediction problem, and it has received little attention until now.

One of the major challenges related to structural SNA that has received a great deal of attention is the problem of community detection, which represents the existence of groups of densely connected nodes. For example, employees working on the same projects in an organization may communicate frequently; people calling each other frequently in a telecommunication network are usually friends, etc. Discovery of such groups, clusters, or communities in a social network forms the basis for many other social network analysis tasks. Community detection in social networks is a challenging task and it is often viewed from different aspects as listed below.

- Local and global community detection
- Detection of overlapping communities
- Detection of hierarchical community structures
- Utilization of link weights and directions
- Community evolution tracking in dynamic networks

1.2   Non-structural Social Network Analysis

Non-structural social network analysis is the domain that involves analytical deductions and pattern recognition primarily from the *user-generated contents* (including texts, images, videos, etc., in which the structural or topological notions are vague) in OSNs. Traditionally, content analysis refers to the systematic review of a body of texts, images, and other symbolic matters with the aim of making valid inferences regarding their contextual usability [17]. Content analysis has been used as a research method for both qualitative and quantitative analyses of information in many fields in which the main task is to extract new facts, patterns, information, and to determine their application domain [17]. The OSNs can be viewed as unstructured-content oceans in which the process of generating new user content in the form of blogs, posts, comments, tags, messages, chats, etc., never stops. We have seen how sometimes user-initiated content storms occur in OSNs and cause revolutions (e.g., the 2011 Egyptian revolution). This calls for methods and techniques to be developed for analyzing user-generated textual contents, which can also help to deal with the challenges discussed earlier, i.e., link prediction (users who generate or consume similar or related content concerning specific topics are more likely to qualify for a missing link or a link in the future), community detection (group of content/textual objects that are related to the same topic subject domain or group of individuals generating and consuming/propagating similar/related content).

With the popularity of the Web and its services, opinions regarding products and services are often expressed by people through formal or informal textual representations using e-mails, online reviews, and blogs. Consequently, the area of opinion mining and sentiment analysis, which is a mutual collaboration of natural language processing, computational linguistics, and text mining, has emerged for the automatic identification and extraction of attitudes, opinions and sentiments from the voluminous amounts of user-generated contents. The motivation here is mainly driven by the commercial demand for cheap, detailed, and timely customer feedback to business organizations. Recent interest in developing such automated systems has swelled to assist the information analysts of various organizations with decision-making and answering questions such as: "how do people feel about the latest camera phone?" In the broader view of Liu et al. [19], the task of opinion mining can be stated as determining the orientation of a sentiment related to a set of objects, expressed in terms of opinions on a set of attributes defining the objects, by a set of opinion holders. According to Liu et al. [19] some important technical challenges related to opinion mining in OSNs are:

- Object identification: The basic requirement is to identify the relevant object that forms a subject for an opinion holder to express opinion in a statement.
- Feature extraction: This involves identifying the object attributes about which opinion holders express their opinions. Sometimes the task is more challenging, as an opinion holder may use alternative words or phrases to refer to a particular

feature of an object in a subjective statement. This requires groups of synonyms describing a particular feature to be found.

- Polarity determination: This involves the subjectivity classification of a statement, as discussed earlier, to determine if a statement contains an opinion and if an opinion exists, thereby also determining its polarity.
- Integration: This highlights the challenge of mapping the challenges mentioned above with each other, i.e., mapping together the orientation of an opinion, projected by the features extracted about a particular object, expressed by a particular opinion holder. It is challenging, as not all information in a statement is explicit.

1.3   Hybrid Social Network Analysis

With hybrid social network analysis, we aim to address all those processes and frameworks that utilize both structural and content-related information of OSNs to extract useful patterns and make meaningful deductions. Some of the most popular hybrid SNA tasks that have received much of the attention are explained in the following paragraphs:

Information Diffusion: The area of information diffusion deals with the analysis and modeling of information flow in social networks with the aim of explaining the spreading processes of various entities such as news, diseases, computer viruses, etc., in the real social world. With the advent of modern communication tools such as instant messaging apps (WhatsApp, Skype), OSNs (Facebook, Twitter), etc., efficient measurement of social contagion is proposed by many researchers, resulting in effective models for information diffusion. Significantly, the works of Katona et al. [15] and Bakshy et al. [1] have thrown some light on the importance of the topological social network properties discussed earlier in this section, and the user-generated content for modeling and analyzing the information diffusion in OSNs.

Influence Analysis: Certain business processes, mainly marketing, are aimed at analyzing the influence of an individual over a population with the goal of utilizing highly influential people to endorse products. For example, in line with the spread of computer or pathological viruses through self-replicating process, viral marketing exploits social contagion and influence ranking to promote brand awareness or to advertise. In the literature, influence analysis has been mostly performed following a hybrid model, i.e., by incorporating both the content and the network structure for analysis.

Semantic Analysis: The amalgamation of the Semantic Web framework and the social media platform is looked upon as the next step toward constructing collective intelligence and knowledge systems for effective information retrieval and efficient content search [12]. One of the most popular ontologies used to represent the network of people is the *friend-of-a-friend* (FOAF), which describes people, their relationships, and actions. Current state-of-the-art related to the Semantic Web attempts automatic construction of vocabulary models from knowledge sources existing in the form of natural texts, based on the observation that OSN users create contextual ontologies through collective intelligence [20]. Such ontologies can aid in the annotation of user-generated content within their communities based on the fact that OSN services also allow users to annotate their published resources using short descriptive tags.

Recommender Systems: Recommender systems are special information filtering systems that are aimed at reducing the domain of object instances (books, videos, friends, news, events, research papers, etc.) based on the level of interest (automatically determined or explicitly mentioned) shown by users toward them. Introduced in the mid-1990s and based on the concept of collaborative filtering, the recommendation problem is formulated as the problem of estimating ratings (predicting the level of interest a user can show toward a particular item) for the items that have not been seen by the users [14]. The problem of designing recommender systems is aimed at developing systems and techniques based on the concept of social networks that assist and augment the natural social process of recommendations along with determining resources (previously unseen) that may match an individual's interests. Various problems related to exploiting OSNs for effective recommendations have been identified by the researchers in [13]. Some of them are briefly described in the following paragraphs:

- How can the merger of social networks and recommendations help to increase consistent user participation and contribution? Is it possible to encourage like-minded users through user-matching via collaborative filtering [8]?
- Is it possible to increase trust in recommender systems by utilizing the various aspects of social networks such as published friend relations in OSNs (preferred recommenders)? [8]? Do friends share preferences, and if they do, how can such information be exploited by recommender systems [13]?
- Is it possible to tackle the cold-start problem by explicitly specifying the users' closest neighbors [8]?
- How weak and strong ties among users can be useful for search and recommendation purposes?

The field of SNA is vast and we would like to stress that a single book or paper cannot do it justice. In this regard, this chapter emphasizes community detection and presents a unified framework that is aimed at facilitating other SNA-related tasks.

2 Community Detection

The task of community detection involves identifying groups (communities) through the study of network structures and topologies. As discussed in Sect. 1.1, community detection is highly challenging, as it involves numerous aspects, which are briefly summarized in the following paragraphs.

Overlapping and hierarchical communities: Real-world community structures show overlapping behavior, i.e., nodes may belong to multiple communities. For example, in a social network, a person may be a member of multiple work groups, and thus it may be significant to identify all such overlapping memberships. In the literature, however, most of the community detection approaches tend to find mutually exclusive community structures. It is thus highly desirable to devise approaches for identifying overlapping community structures. Besides showing overlapping behavior, community structures may possess a hierarchical structure at different resolutions in which a larger community contains multiple smaller communities. Very few methods described in the literature address these two issues collectively, leading to the need for further research to fill this gap.

Dynamic communities and their evolution: An important property of the real-world social networks lies in their tendency to change dynamically, i.e., the addition/deletion of new nodes and links, and the change in the intensity of interactions. These changes directly affect the structure of underlying communities synchronously resulting in *birth*, *growth*, *contraction*, *merge*, *split*, and *death* of communities over time. Some of the methods that have been proposed to identify the evolution of communities with time include [11, 26], and [16]. However, Lin et al. [18] pointed out that these methods have a common limitation in which communities and their evolution have been studied separately.

Utilities: Communities in social networks map to functional modules such as functionally coherent groups of proteins in protein interaction networks, discussion groups in OSNs, etc. The community structure of a network can also be used to determine the roles of individual members, in which boundary nodes can be used to facilitate interactions between two or more community members, whereas the central nodes may play an important role in the control and stability of the community. Knowing an underlying community structure of a communication network can be used for spammer identification in OSNs by characterizing the interaction behavior of legitimate users within the identified communities. Spammer interaction can then be filtered out or blocked based on the source and target communities of the interaction.

3   A Unified Community Analysis Framework

We have presented a unified community detection and evolution analysis framework, HOCTracker, in [7], which is derived from a density-based clustering approach, in which a cluster is searched by detecting the neighborhood of each object in the underlying database. For any node, its neighborhood is defined in terms of its distance from other nodes in the database. A node q is assigned to the neighborhood of another node p only if the distance between them is less than or equal to a threshold ε. In a network context, this distance constraint (which is usually a structural metric) is checked only for those nodes that have a link/edge between them. Depending upon whether the neighborhood of a node contains more than μ nodes, a new cluster with p as a core object is formed. Building upon the notion of these core nodes, a density-connected cluster is identified as a maximal set of density-connected nodes. Density-based clustering and community detection methods are relatively fast and easily scalable to large databases/networks with simplified parallel implementations. However, density-based methods require two input parameters that include the global neighborhood threshold, ε, and the minimum cluster size, μ toward which they are highly sensitive. Estimating an optimal value for the ε parameter automatically is a long-standing challenge [24]. HOCTracker, on the other hand, automatically estimates the value for ε based on the local neighborhood of each node. A similar approach followed for μ uses a resolution parameter η tuned as required or estimated using the heuristic approach given in [7].

Figure 2 presents a conceptual overview of the proposed social network analysis framework, which consists of three overlapping but inter-related functioning modules, each addressing one particular aspect of community analysis: *overlapping community detection*, *community evolution tracking*, and *hierarchical community structure identification*.

The core constituent of all these modules is the identification of density-based structural components. For a given value of the resolution parameter (η), the basic overlapping community detection process involves identifying overlapping density-based components from the given state of a network. It follows an iterative and local expansion approach to directly expand a community from a randomly selected seed

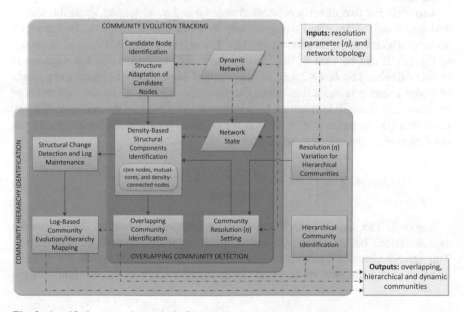

Fig. 2 A unified community analysis framework

node until all nodes in the network are visited. It allows a node to be a member of multiple communities and thus identifies overlapping communities for a social network.

The evolution detection and tracking module adapts a reference community structure, identified at any initial stage of a dynamic network, to the changes occurring in the underlying network. This involves the reprocessing of only active node neighborhoods in the network, which is different from other methods that require reprocessing neighborhoods of all nodes. The mapping of evolutionary relationships between communities across consecutive time-steps is simplified by using an efficient log-based approach.

The hierarchical community detection module is aimed at generalizing the overlapping community detection and evolution tracking process to identify and map the hierarchical relations between overlapping communities identified at varying resolutions by finding density-based structural components at different resolution parameter values (η).

3.1   Distance Function

Although the complete details of the method can be seen in our research paper [7], here we briefly present the novel dual-layer distance function used by HOCTracker, incorporating both directed and weighted attributes of the edges, if available. A brief description of the distance function is given in the following paragraphs.

Layer-1: For two directly connected nodes p and q, let V_p and V_q be the sets of nodes to which nodes p and q respectively have out-links, and let V_{pq} be the set of nodes to which both p and q have out-links. The *layer-1* distance is defined as shown in Eq. (1). It can be seen in Eq. (1) that the function ensures that the distance is further calculated by *layer-2*, i.e., $d(p,q)$, only if the number of common recipients of nodes p and q is more than some fraction (specified by η) of the minimum of the two. Otherwise, it is taken as 1 (maximum). At this point we can note that the fraction η (in the range $0 < \eta \le 1$) is the required input parameter to specify the resolution of communities to be identified.

$$\text{dist}(p,q) = \begin{cases} d(p,q) & \text{if } |V_{pq}| > (\eta \times \min(|V_p|, |V_q|)) - 1 \\ 1 & \text{otherwise} \end{cases} \tag{1}$$

Layer-2: The second layer of the distance function is based on the reciprocity of interactions between the two nodes, including their common neighborhoods. The *layer-2* distance function is given by Eq. (2) in which $I_{\overleftrightarrow{pq}}$ is the number of reciprocated interactions between nodes p and q taken as the minimum of the interactions from p to q and vice versa.

$$\delta(p,q) = \begin{cases} \left(\frac{\sum_{s \in V_{pq}}(I_{\overleftrightarrow{ps}}) + I_{\overleftrightarrow{pq}}}{|V_{pq}|+1} \right) & \text{if } I_{\overleftrightarrow{pq}} > 0 \\ 0 & \text{otherwise} \end{cases} \tag{2}$$

To obtain a symmetrical distance, $d(p,q)$ is taken as the maximum of their mutual directed response (*or minimum of the reciprocals of their mutual directed response*) values normalized by their respective total weight of outgoing interactions (represented by $I_{\overrightarrow{p}}$ and $I_{\overrightarrow{q}}$ respectively) in the interaction graph, as given in Eq. (3).

$$d(p,q) = \begin{cases} \min \left(\frac{\delta(p,q)^{-1}}{I_{\overrightarrow{p}}}, \frac{\delta(q,p)^{-1}}{I_{\overrightarrow{q}}} \right) & \text{if } \delta(p,q) > 0 \wedge \delta(q,p) > 0 \\ 1 & \text{otherwise} \end{cases} \tag{3}$$

The dual-layer distance function thus measures the maximum average reciprocity between two nodes and their common neighbors only if the overlap of their neighbors is significant (specified by η). Smaller values for $d(p,q)$ represent a higher response between nodes p and q and translates to more closeness between p and q (less closeness for higher values).

As the proposed method adapts a density-based approach, a neighborhood threshold needs to be specified to mark the boundary of the smallest cluster possible. Unlike other density-based methods, HOCTracker automatically determines (for any node) a local version of the neighborhood threshold, called the *local-neighborhood threshold* represented by ε. For a node p, the value for ε_p is calculated as the average of reciprocated interactions of p with all its out-linked neighbors and defined using Eq. (4).

$$\varepsilon_p = \begin{cases} \frac{\left(\frac{I_{\overleftrightarrow{p}}}{|V_p|} \right)^{-1}}{I_{\overrightarrow{p}}} & \text{if } |V_p| > 0 \wedge I_{\overleftrightarrow{p}} > 0 \\ 0 & \text{otherwise} \end{cases} \tag{4}$$

In the above equation, $I_{\overleftrightarrow{p}}$ is the amount of reciprocated interactions for node p (i.e., $\sum_{\forall q \in V_p} \min(I_{\overrightarrow{pq}}, I_{\overrightarrow{qp}})$), and $\frac{I_{\overleftrightarrow{p}}}{|V_p|}$ is the average reciprocated interaction between node p and all other nodes in V to which p has out-links. The denominator $I_{\overrightarrow{p}}$ normalizes the value of ε_p within the range $[0, 1]$ and represents the total out-link weights of node p in the interaction graph of the social network.

Based on the above definitions of distance function dist(p,q) and threshold ε_p, the *local ε_p-neighborhood* of a node p, represented by N_p, consists of a subset of V_p with which the distance of p is less than or equal to ε_p, and it is defined formally by Eq. (5).

$$N_p = \{q : q \in V_p \wedge \text{dist}(p,q) \leq \varepsilon_p\} \tag{5}$$

To define the notion of a density-based community in a social network at a given resolution fraction (η), the proposed approach uses the following key definition.

Definition 1 (Core Node) A node p with non-zero reciprocated interactions with any of its neighbor(s) in V_p is designated as a core node if its local ε_p-neighborhood contains at least μ_p (local minimum-number-of-points threshold for p) of nodes in V_p, as given in Eq. (6), where $\mu_p = \eta \times |V_p|$.

$$\text{CORE}_\eta(p) \Leftrightarrow |N_p| \geq \mu_p \tag{6}$$

3.2 Community Detection and Tracking

To start the process of community detection, HOCTracker selects a random node p and computes the local ε_p threshold using Eq. (4). If the value of ε_p is greater than zero, the dual-layer distance function dist(p, q) mentioned in Sect. 3.1 is used to identify the N_p of p. Thereafter, depending upon the result of the core node test given in Eq. (6), the following procedure is used to find the communities:

1. If node p forms a core node, the set of visited nodes V with which p has mutual core relations and the set U of un-visited nodes in N_p are determined.
2. If V is empty, a new community is formed by including p and all nodes in N_p. Moreover, the community memberships of all the nodes in this newly formed community are appended with a new community ID C, and node p forms a primary core of community C. If N_p includes some core nodes that are not in the set V, they form the secondary core nodes of C.
3. If V is non-empty, and all the core nodes in V are primary cores of a single community C, then p joins and also forms a primary core of C. The community memberships of all the nodes in N_p, including p, are appended with label C.
4. If V is non-empty and some core nodes in V are primary cores of different communities, then the communities are merged to form a single community. A new community ID replaces the community membership of the merged communities for all visited density-reachable nodes of p, and the primary nodes of merged communities including p form the primary nodes of the new community.
5. Node p is marked as visited.
6. For each node q in U, if q is not marked as *waiting* then q is marked as *waiting* and q added to the queue.
7. Repeat these steps for each node removed from the queue until the queue is empty.

If the randomly selected node p does not qualify as a core, it is marked visited and may be added as a non-core to the neighborhood of some other core. Otherwise, it is treated as an outlier. The above process is iteratively repeated for each new randomly selected node until all nodes in the network are visited. This process finally identifies an overlapping community structure from an underlying network.

To track and map the evolution of communities in a dynamic network at two successive time-steps, HOCTracker uses a log called *intermediate evolution log (IEL)* to record the intermediate transitions occurring in the communities after processing the neighborhoods of active nodes. All intermediate community transitions caused by re-computing the local ε-neighborhood of candidate nodes at a new time-step result in entry into the IEL. The information thus stored in the log is then used to efficiently deduce a bipartite mapping between the communities at different time-steps.

We have already mentioned that HOCTracker requires a single parameter η to be specified for identifying communities at a particular resolution. The value of η, although inversely proportional to the size of the communities identified, can be easily tuned to detect communities at different levels of size characteristic, forming a hierarchical representation of overlapping community structures. To establish hierarchical relationships between communities identified at different levels (different values of η) HOCTracker uses the same log-based evolutionary mapping technique discussed earlier. Communities at a higher level can be viewed as resulting from the *merge* and *growth* of communities at lower levels. Similarly, community structure at a lower level can be considered to result from the *death*, *split*, and *shrinkage* of communities at a higher level. This forms a unified framework for detecting and tracking overlapping and hierarchical communities in dynamic social networks.

For a dynamic network, three values of η are estimated for each new state of a network by following a heuristic. Before community structure is identified for a new level, the community structure and state of evolution log for the previous level are saved. If the modularity score for the next level is lower than for the previous level, the saved state is reloaded to represent communities for the current state of the network and is used for mapping evolutionary relations. A simple illustration of community detection and tracking by HOCTracker on a dynamic network involving three time-steps is shown in Fig. 3.

4 Applications of Community Structures

As mentioned earlier, community structure within a network represents the existence of groups of densely connected nodes with only sparse connections between the groups. For example, employees who work on related projects in an organization may communicate frequently via email; people who call each other more often in a telecommunication network are usually friends, etc. Discovery of such groups, clusters, or communities in a social network forms the basis for many other social network analysis tasks and finds its application in numerous domains. For example, communities in social networks map to functional modules such as functionally coherent groups of proteins in protein interaction networks, discussion groups in

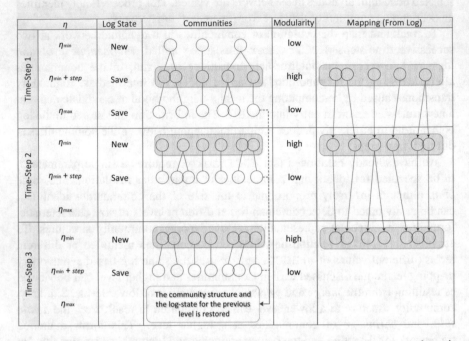

Fig. 3 An illustration of community detection and tracking by HOCTracker on a dynamic network with three time-steps

OSNs, etc. However, not many research works deal with the issue of utilization of detected communities and simply propose a method of detecting them without emphasizing what to do with them later. In this section, we present some novel application domains and methods for the utilization of communities detected by community detection methods. These application domains include spammer detection in OSNs, detection of copy profiling attacks, and identifying influential nodes for viral marketing.

4.1 Online Spammer Detection

One of the most spiteful activities performed in OSNs is spamming, where a spammer broadcasts his/her desired content to as many benign users as possible through e-mails, instant messages, comments, posts, etc. The main aim of such an activity is the promotion of products, viral marketing, spreading fads and fake news, and sometimes to harass other users. To tackle the nuance of spamming, devising methods for the automatic detection of spammers and their spamming activity is a highly desirable task. Detected spammers can be removed from social networks, and future spamming activities can be significantly controlled by analyzing spammers' behavior and taking precautionary measures.

In the literature, numerous spammer detection methods have been proposed using content analysis, mainly involving keywords-based filtering. However, spammers have developed many counter-filtering strategies incorporating non-dictionary words and images in the spam content. The most common limitations of content-based spam filtering systems include the need for a high number of computations and the issue of users' privacy. The need for access to users' private messages, posts, profile details, etc., is often considered a negative aspect for the content-based spam-filtering systems. Recently, spammer detection methods have utilized classification models for learning rules from social network-based topological features, including indegree, outdegree, reciprocity, clustering coefficient, etc. [25]. Along these lines, Bhat and Abulaish [4] proposed a community-based spammer detection method for OSNs by proposing novel community-based node features that can be extracted from the overlapping community structure of the underlying OSN. For each node, the community-based features that they are aimed at extracting include the following:

- Whether or not the node is a *core-node* (as discussed in the previous section).
- The number of participating communities for a node.
- The number of nodes belonging to other communities to which it has out-links.
- The ratio between the number of its in-links from outside of its community to that of its out-links.
- The probability of a node having an out-link outside of its community.
- The reciprocity ratio with the nodes outside of its community.
- The probability that the non-member out-linked nodes belong to a common foreign community.

Based on these features, Bhat and Abulaish [4] proposed a classification framework to identify the spammers in OSNs, as shown in Fig. 4. The evaluation of the proposed spammer classification system over Facebook and Enron datasets reveal that the novel community-based features are significant for identifying spammers in OSNs.

4.2 Detecting Copy-Profiling/Cloning Attacks

One of the most deceptive forms of malicious attacks seen over OSN platforms nowadays is the *cloning attack*, which is different from traditional spammer attacks and is often difficult to detect and track. A cloning attack (aka copy-profiling attack) involves an attacker creating clones (similar accounts) of other benign accounts (targets) by copying and replicating their profile details. Then the attacker sends friend requests to some of the friend accounts of the target account with a hope that some non-suspicious account users may be deceived and accept the friend request from the clone. The deception becomes possible because some recipients consider

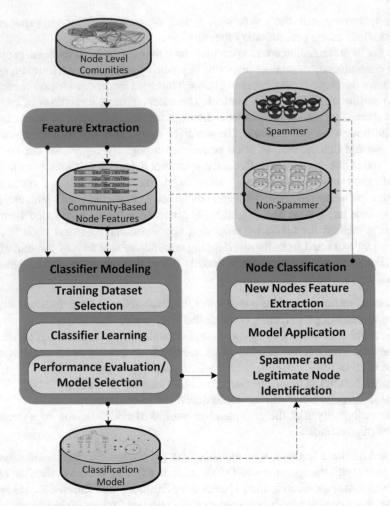

Fig. 4 Spammer classification model learning and application

the friend request to be coming from the actual benign user cloned by the attacker, without realizing that a similar account is already befriended by them. This initial befriending with some benign accounts forms the first stage of the cloning attack. After this stage, it becomes easier to breach the social circle and trust of users, who now share friends with the clone [6]. Figure 5 illustrates a two-stage clone. Here, Fig. 5a shows the first stage, in which an attacker clones a benign user (Ben) and multicasts friend requests to some of the Ben's friends, a few of whom accept.

The second stage of the attack is illustrated in Fig. 5b, in which the attacker selects some of those accounts with which the clone has common friends. To maintain stealth and increase the level of infiltration, a cloning attack in an advanced form may involve creating multiple clones of different accounts at each subsequent infiltration stage. Once an attacker has infiltrated to a significant level

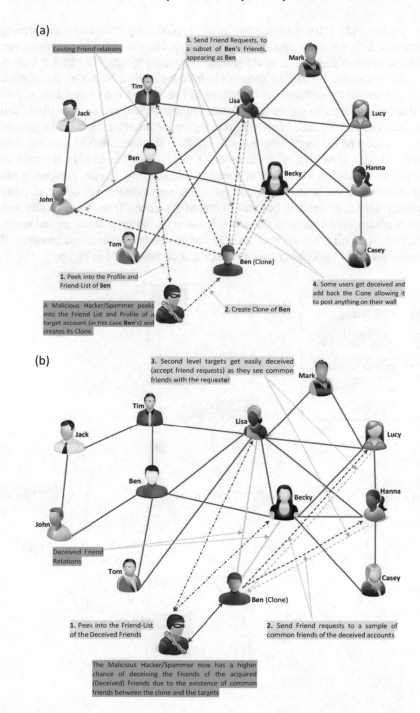

Fig. 5 Illustration of a cloning attack in two stages

of its requirements, it has multiple options to exploit the privileges by performing malicious activities such as posting spam messages, launching phishing attacks, etc.

For each cloned account, an attacker follows a greedy strategy in which it tries to maximize the number of benign accounts befriended while minimizing the number of friend requests broadcasted as illustrated in Fig. 5. In a bid to avoid detection, the attacker also tries to gather a moderate number of friends for each clone and links these clones with each other to mimic legitimate behavior. However, a general behavior observed for legitimate users in OSNs is the formation of communities whereby similar users tend to form separate overlapping clusters. In the case of dynamic networks, the frequency of merging and splitting of communities is not frequent, whereas the birth, death, and growth of communities are observed more frequently. This behavior is in contrast to that of the cloned accounts of an attacker, which result in dense regions in the network due to the addition of clones and cross-links. Robust cloning behavior of an attacker results in the frequent merging of communities in the underlying social network, as demonstrated in Fig. 6.

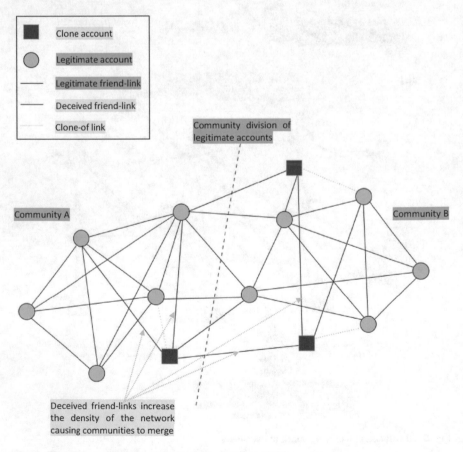

Fig. 6 Illustration of the robust cloning behavior of an attacker

Based on the above observation, it can be argued that a community detection and tracking method such as that discussed in Sect. 3 can detect and track cloning attacks by detecting regions facing frequent community merge events in the underlying network. Such regions can be explored by conducting profile similarity tests and neighborhood overlap analysis, and those with a high level of similarity can be considered a clone.

4.3 Influential Node Detection for Viral Marketing

Viral marketing (VM) is a strategy of exploiting a self-replicating process such as that of viruses, with the aim of increasing product/brand visibility and product sales [5]. The main process involves identifying a few nodes in a social network that have a great influence and developing promotion strategies for that smaller set. Thus, one of the preliminary tasks of a viral marketing campaign is the identification of influential nodes, and this is generally a challenging task. In this regard, Bhat and Abulaish [5] suggested that the nodes that are shared by overlapping communities might be good candidates to be considered influential. In addition, density-based community detection methods, as discussed in Sect. 3, identify hubs that have similar properties to that of overlapping nodes and can also form good seed nodes for a VM campaign. Based on these assumptions, Bhat and Abulaish [5] tested their hypothesis on an email dataset and argued that overlapping nodes are among the best candidates to be considered as influential nodes. As shown in Fig. 7, their analysis reveals that an overlapping node has a higher chance of appearing in the

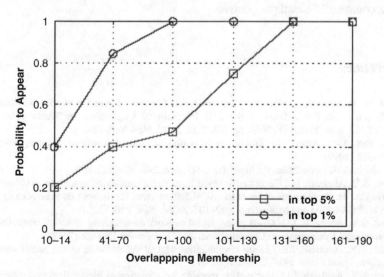

Fig. 7 Probability distribution of overlapping nodes to appear in the top influential 1% and 5% of the total nodes

top 1% and 5% of influential nodes in a social network, ranked according to their betweenness centrality, thus, strengthening the argument that overlapping nodes are highly influential in social networks.

5 Conclusion and Future Work

In this chapter, we have presented a unified community analysis framework with the aim of addressing the major challenges, including the detection of overlapping and hierarchical communities, and tracking their evolution in dynamic networks. Unlike many other research works in this field, we have also presented some of the important application areas for the identified communities, including detection of spammers and cloning attacks and leveraging community structures for the task of viral marketing. One of the future directions of research in this field may be the realization of a multi-dimensional framework that utilizes both structural and non-structural information for social network analysis tasks, for example, to analyze the topological community structures and their evolution in the light of topical user sentiments on a common ground, to generate greater insights related to information diffusion in a social network. Some important insights that may be gained by merging structural communities and user-centered topics/sentiments include identification of communities around discussion topics/sentiments and vice versa, analyzing the correlation between the dynamics of communities (birth, merge, split, etc.) and the dynamics of discussion topics and sentiments, identifying opinion leaders within communities, diffusion of information (topics, rumors, fads, news) within and across communities, and modeling the information diffusion process from a community-based perspective.

References

1. Bakshy, E., Rosenn, I., Marlow, C., Adamic, L.: The role of social networks in information diffusion. In: Proceedings of the 21st International Conference on World Wide Web, WWW'12, New York, NY, 2012, pp. 519–528. ACM, New York (2012)
2. Barabási, A.-L., Albert, R.: Emergence of scaling in random networks. Science 286(5439), 509–512 (1999)
3. Barabási, A.-L., Bonabeau, E.: Scale-free. Sci. Am. 288, 50–59 (2003)
4. Bhat, S.Y., Abulaish, M.: Community-based features for identifying spammers in online social networks. In: Proceedings of the IEEE/ACM International Conference on Advances in Social Networks Analysis and Mining, pp. 100–107. ACM, New York (2013)
5. Bhat, S.Y., Abulaish, M.: Overlapping social network communities and viral marketing. In: International Symposium on Computational and Business Intelligence (2013)
6. Bhat, S.Y., Abulaish, M.: Using communities against deception in online social networks. Comput. Fraud Secur. 2014(2), 8–16 (2014)
7. Bhat, S.Y., Abulaish, M.: Hoctracker: tracking the evolution of hierarchical and overlapping communities in dynamic social networks. IEEE Trans. Knowl. Data Eng. 27(4), 1019–1013 (2015)

8. Bonhard, P.: Who do trust? Combining recommender systems and social networking for better advice. In: Proceedings of the Workshop on Beyond Personalization, in Conjunction with the International Conference on Intelligent User Interfaces, pp. 89–90. Citeseer (2005)
9. Getoor, L.: Link mining: a new data mining challenge. SIGKDD Explor. **4**, 1–6 (2003)
10. Granovetter, M.: The strength of weak ties. Am. J. Sociol. **78**(6), 1360–1380 (1973)
11. Greene, D., Doyle, D., Cunningham, P.: Tracking the evolution of communities in dynamic social networks. In: Proceedings of the International Conference on Advances in Social Networks Analysis and Mining, ASONAM '10, Washington, DC, pp. 176–183. IEEE Computer Society, New York (2010)
12. Gruber, T.R.: Collective knowledge systems: where the social web meets the semantic web. Semantic Web **6**(1), 4–13 (2008)
13. He, J., Chu, W.W.: A Social Network-Based Recommender System (SNRS). Springer, New York (2010)
14. Hill, W., Stead, L., Rosenstein, M., Furnas, G.: Recommending and evaluating choices in a virtual community of use. In: Proceedings of the SIGCHI Conference on Human Factors in Computing Systems, pp. 194–201. ACM Press/Addison-Wesley Publishing Co., New York (1995)
15. Katona, Z., Zubcsek, P.P., Sarvary, M.: Network effects and personal influences: the diffusion of an online social network. J. Mark. Res. **48**(3), 425–443 (2011)
16. Kim, M.-S., Han, J.: A particle-and-density based evolutionary clustering method for dynamic networks. Proc. VLDB Endowment **2**(1), 622–633 (2009)
17. Krippendorff, K.: Content Analysis: An Introduction to Its Methodology. Sage, Thousand Oaks (2004)
18. Lin, Y.-R., Chi, Y., Zhu, S., Sundaram, H., Tseng, B.L.: Analyzing communities and their evolutions in dynamic social networks. ACM Trans. Knowl. Discov. Data **3**, 8:1–8:31 (2009)
19. Liu, B., Chen-Chuan-Chang, K.: Editorial: special issue on web content mining. ACM SIGKDD Explor. **6**(2), 1–4 (2004)
20. Mika, P.: Social networks and the semantic web: the next challenge. IEEE Intell. Syst. **20**(1) (2005)
21. Milgram, S.: The small world problem. Psychol. Today **2**(1), 60–67 (1967)
22. Moreno, J.L.: Who Shall Survive? A New Approach to the Problem of Human Interrelations. Nervous and Mental Disease Publishing Co., Washington, DC (1934)
23. Newman, M.E.J., Park, J.: Why social networks are different from other types of networks. Phys. Rev. E **68**(3), 036122 (2003)
24. Sun, H., Huang, J., Han, J., Deng, H., Zhao, P., Feng, B.: gSkeletonClu: density-based network clustering via structure-connected tree division or agglomeration. In: Proceedings of the 2010 IEEE International Conference on Data Mining, ICDM'10, Washington, DC, pp. 481–490. IEEE Computer Society, Washington (2010)
25. Wang, A.H.: Don't follow me: spam detection in twitter. In: Proceedings of the 2010 International Conference on Security and Cryptography (SECRYPT), pp. 1–10 (2010)
26. Wang, Y., Wu, B., Du, N.: Community Evolution of Social Network: Feature, Algorithm and Model. arXiv: physics.soc-ph (2008)
27. Wasserman, S., Faust, K.: Social Network Analysis: Methods and Applications, vol. 8. Cambridge University Press, Cambridge (1994)
28. White, H., Boorman, S., Breiger, R.: Social structure from multiple networks: I. Blockmodels of roles and positions. Am. J. Sociol. **81**(4), 730–780 (1976)

GSO Based Heuristics for Identification of Communities and Their Leaders

Nidhi Arora and Hema Banati

Abstract Community identification plays an important role in classifying highly cohesive functional subunits from a larger complex networked environment. Central nodes in these communities are primary agents for regulating information flow locally within and globally across the communities. The significant contribution of this chapter is a novel metric to identify community-specific central nodes (leaders) that are also highly trusted globally in the overall network. This paper uses a very efficient nature-based DTL-GSO heuristic optimization algorithm based on the Group Search Optimization (GSO) algorithm for the extraction of communities by following a framework-based approach. Using the proposed approach, designated communities and their leaders can facilitate the task of harmonizing community-specific assignments in a more efficient manner, leading to an overall performance boost. This chapter also contains a thorough discussion of the many variants of GSO based community detection algorithms and can thus be used as a comprehensive guide to model any new nature-based algorithm for this problem domain. Detailed experimental results using real and synthetic relationship networks are presented, followed by a real-life case study of a school children relationship network.

Keywords Centrality • Community detection • Group search optimization • Modularity • Relationship networks

1 Introduction

Understanding real-world relationships or interactions to analyze complex interconnections among real world entities has been a prominent research area in varied field of disciplines. Modelling these complex interconnections and entities

N. Arora (✉)
Kalindi College, University of Delhi, New Delhi, India
e-mail: nidhiarora@kalindi.du.ac.in

H. Banati
Dyal Singh College, University of Delhi, New Delhi, India
e-mail: hemabanati@dsc.du.ac.in

© Springer International Publishing AG 2017
H. Banati et al. (eds.), *Hybrid Intelligence for Social Networks*,
DOI 10.1007/978-3-319-65139-2_5

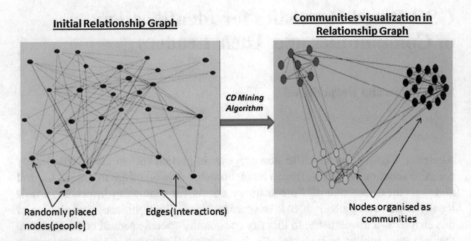

Fig. 1 Community visualization in relationship networks

in the form of relationship networks/graphs have resulted in the application of graph and network theory concepts for their analysis. Entities in these relationship networks are represented as nodes and their interactions as edges, as shown in Fig. 1. The presence of communities is a very significant property of such networks. Communities are densely connected subsets of nodes such that the overall density of edges within the subsets is greater than the overall density of edges across the subsets [33]. These communities are subgroups of highly interactive nodes, which, when identified, can help to make strategic level decisions for improving productivity. Communities are not directly visible as real-world social networks can be too complex owing to their dense connectivity; hence, appropriate community detection (CD) algorithms are applied to mine perfectly connected subdivisions of nodes out of these networks. Figure 1 shows how the application of a chosen CD algorithm on a sample network has resulted in partitioning of the network into three distinct communities.

This chapter is organized as follows: a literature review is given in Sect. 2 and related definitions are given in Sect. 3, followed by a generalised, nature-based CD framework in Sect. 4; Sect. 5 introduces a basic group search optimization (GSO) algorithm and its modelling to a CD problem; Sect. 6 discusses variants of GSO for a CD problem in detail, followed by the introduction of extracting community-wise central nodes in Sect. 7; and a real-life case study of a school children relationship network is presented in Sect. 8.

2 Literature Review

It is computationally expensive (NP hard) to extract perfect and optimal partitioning called communities of nodes using standard polynomial time algorithms [16, 41]. Hence, CD algorithms are directed by heuristics or greedy selection approaches to optimize some community-specific quality function and extract near to optimal communities. Many community quality functions are described in the literature [43] that can be chosen for optimization in CD algorithms. These functions generally consider the strength of intra-community vs inter-community edges to quantify community quality. Modularity metric [33] is an example of a commonly optimized community quality function. GN [23], fast GN [33] and CNM [17] are some of the initially defined CD algorithms that apply greedy techniques to optimize modularity objective function.

The optimality of extracted communities may vary depending upon the heuristics being applied in the algorithm. Nature-based heuristic techniques have recently gained much significance because of their capability of providing acceptably optimal solutions to many complex and computationally hard problems in various application areas of computing [19, 45]. This field of computing stochastically exploits the intelligence hidden in the nature for selection of the fittest, evolution, recognition, understanding and learning, while handling uncertainty, to achieve acceptable solutions for complex problems. Recently, these techniques have also been tested to provide optimal solutions to CD problems. Nature-based genetic algorithm with a local search was applied by Pizzuti [37] to detect modular community structures; Gach and Hao [22] applied the memetic algorithm to detect communities in complex networks; Hafez et al. [26] used the artificial bee colony (ABC) swarm optimization algorithm for the detection of communities; Amiri et al. [4] applied the firefly algorithm to detect communities. Many other approaches [2, 5, 8, 15, 25, 47] use various recently proposed nature-based algorithms to optimize chosen community quality metric/metrics and extract the optimal communities.

This chapter presents a comprehensive guide to model a chosen nature-based approach for detecting communities by considering an efficient nature-based GSO algorithm [27] as a model algorithm. Modelling of GSO algorithm for detecting communities is presented using a framework-based approach. The GSO algorithm is chosen because of its efficient population optimization strategies using small-sized populations. Many heuristically chosen domain-specific modifications in the basic GSO algorithm are also presented in detail as variants of GSO to improve its optimality and convergence by utilizing many different domain-specific aspects of complex networks independently or in combination. The GSO variants discussed are: the Mod-GSO evolutionary algorithm [8]; the multi-objective GSO algorithm [5]; a hybrid TL-GSO algorithm [9]; a hybrid discrete DTL-GSO algorithm [10] and a node similarity-enhanced E-GSO algorithm [6]. These variants can give the reader an insight into building intuition for the consideration of many possible domain aspects in varied directions to intuitively improve the efficiency of one's chosen nature-based algorithm. Later, a novel approach to identify community-specific leaders is proposed; for this, a new metric *node_rank* is proposed that quantifies the node's significance in the community for selecting group-specific leaders.

3 Definitions

Community: A community (or cluster) in a network $G(V, E)$ (having $\{V\}$ as a set of nodes and $\{E\}$ as a set of edges) is defined as a subset c of nodes of V such that the density of edges within c is greater than the density of edges connecting nodes in c to the rest of the network [38]. A relationship network G can be partitioned into k such communities/subsets $(c_1, c_2, c_3.c_k)$. A community c_i is strong if for every node a in c, $ka_{in}(c_i) > ka_{out}(c_i)$, where $ka_{in}(c_i)$ is the number of edges that connects node a to the other nodes within c_i (in degree) and $ka_{out}(c_i)$ is the number of edges that connects a to the rest of the network (outdegree) [4].

Modularity: Modularity (Q) [33] is a community quality measure that quantifies the variance between actual edges among all the pairs of nodes within the community and the edges present those pairs in a randomly created model network of the same network. The randomly created network distributes the edges of the actual network randomly among nodes while maintaining the degree distribution of the original network. Modularity can be calculated for a given community partition by applying Eq. (1).

$$Q = \frac{1}{2m} \sum_{vw} \left[A_{vw} - \frac{(k_v * k_w)}{2m} \right] \delta(c_v, c_w) \tag{1}$$

In the above equation, A_{vw} is the number of edges observed between a pair of nodes v and w in a community c and $(k_v * k_w)/2m$ is the number of edges expected at random between the two in the equivalent random network. Here, m is the total number of edges and k_v and k_w are the degrees of nodes v and w respectively. Delta function is 1 if a nodes v and w belong to the same community and 0 otherwise. Modularity value for a given community partition lies within the range [1/2, 1) with positive values representing good community structures and 0 and negative values representing no community structure [33].

Community detection as an optimization Problem: community detection can be formulated as an optimization problem aimed at partitioning nodes of network G into such a k number of distinct subsets (communities) that best satisfies a quality measure $F(G)$ by working towards optimizing (maximize/minimize) $F(G)$ [26].

4 Nature-Based CD Framework

Nature-based heuristic algorithms use intelligence inspired from nature to optimize a chosen community quality function. This section presents a framework for applying natural intelligence-based optimization strategies to detect communities. The framework can then be used as an abstraction for evaluating any set of new nature-based methodologies for detecting communities at varied levels of optimalities. The framework consists of four main phases: analysis, initialization, evolution and result generation. Phases of this framework are designed to be independent with each other to make it adaptable and easy for users to evaluate multiple algorithms by modifying the evolution phase only (Fig. 2).

Analysis Phase: This phase first identifies key entities(individuals) in an organization that are required to be organized in communities. In this, interactions are recorded for selected entities and then used to generate a relationship network. For n individuals, an n cross n matrix A is created such that every entry $A[i][j]$ contains the number of interactions for ith and jth individuals. Significant interactions are then identified (domain-specific criteria can be used) to create a network of n nodes with edges drawn only between nodes that have significant interactions [7].

Initialization Phase: The generated interaction network is then used to generate a population of candidate community solutions. A locus-based adjacency (LBA) initialization strategy is generally used here to represent individuals in the population [4]. In this representation, an individual solution Y contains a vector y1, y2, ...yN of N variables (N is the number of nodes). Each position of this solution vector represents a node of the considered network. Each node position in this vector is filled with a value picked randomly from the adjacency matrix entries of the respective node. A linear time decoding step to form communities checks the entries of each index in an individual solution and puts them in a single community. An individual considering a ten-node network model is shown in Fig. 3. Here Y [0] = 5 implies that there is an edge between node 0 and node 5 and hence puts them in a single community. If either of the nodes is already present in a previously created community, then a new community is not created, but the node is appended to the

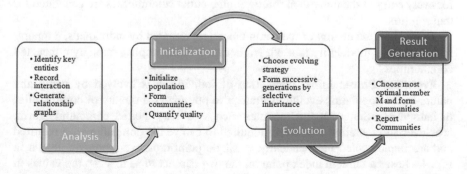

Fig. 2 Framework for nature-based community detection

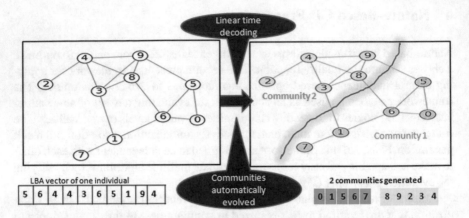

Fig. 3 Locus-based adjacency (LBA) representation and community generation

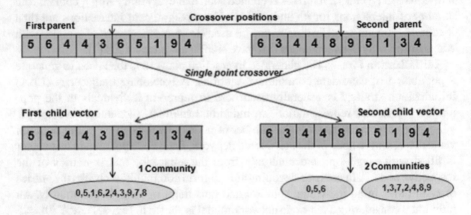

Fig. 4 Single point crossovers between two parent vectors

previously created community containing the other node. This process is repeated for every entry of the individual solution and distinct communities are generated for that solution.

To measure the quality of the communities represented by individuals, a quality metric such as modularity is used. Higher modularity implies strongly connected communities.

Evolution Phase: Initial population of individuals is evolved by using any nature-based population evolution strategy to optimize the quality of communities in individuals. Some support modules such as quality metric, community form, crossover and mutation (which are required to evolve the population) are required and are implemented independently. A single point crossover operator, shown in Fig. 4, choses a random index point on the two parents so as to swap the values in both the vectors to generate two new child vectors. Newly generated child vectors may represent different community partitions with different community quality fitness. The best quality child vectors are only selected in the next generation and the rest are all ignored. The mutation operator can be used to stochastically vary small index positions in a single parent vector to generate a new mutated individual with different fitness (increased or decreased).

Result Generation Phase: This phase then chooses the best member of the last generation of the population and outputs the communities of that member.

5 Group Search Optimization for Community Detection

In this chapter, we model the GSO algorithm as a solution methodology in the evolution phase of the above framework. Basic GSO optimization is discussed below.

5.1 Group Search Optimization (GSO): Algorithm Description

Group search optimization [27] is a nature-inspired, population-based swarm optimization algorithm inspired by animals' group food searching behaviour. Animals in a swarm are placed initially in an n-dimensional space to search for better food locations iteratively (searching bouts). Every ith member at the kth searching bout is assigned a position $X_i^k \epsilon R^n$ and a head angle $\phi_i^k = (\phi_{i1}^k, \ldots \ldots, \phi_{i(n-1)}^k) \epsilon R^{n-1}$. Search direction $D_i^k(\phi_i^k) = (d_{i1}^k, , d_{in}^k) \epsilon R^n$ of ith member is calculated from its head angle via polar to cartesian coordinate transformation [27]. Hierarchies in animal groups are based on the fitness of occupied positions by the animals in the searching space such that the fittest is chosen to be the producer (the leader), about 80% of animals are chosen to be scroungers (followers) and rest as rangers. Phases in the evolution mechanism of GSO are shown in Fig. 5.

Producer X_p is the leader and searches for three new food sources (at 0 degree, right-hand hypercube and left-hand hypercube) in its conical scanning field defined by its head angle and search direction. The scanning field has a maximum pursuit angle and maximum pursuit distance characterized by $\theta_{max} \epsilon R^1$ and $l_{max} \epsilon R^1$ respectively. The three new points are generated by using the following equations.

$$\text{Point at zero degree}: X_z = X_p^k + r_1 * l_{max} * D_p^k(\phi^k) \tag{2}$$

$$\text{Point at right hand hypercube}: X_z = X_p^k + r_1 * l_{max} * D_p^k(\phi^k + r_2\theta_{max}/2) \tag{3}$$

$$\text{Point at left hand hypercube}: X_z = X_p^k + r_1 * l_{max} * D_p^k(\phi^k - r_2\theta_{max}/2) \tag{4}$$

Fig. 5 GSOs population
evolution mechanism

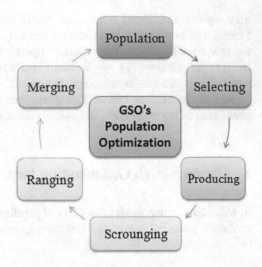

Here, $r_1 \epsilon R^1$ is a normally distributed random number with a mean 0 and standard deviation 1 and $r_2 \epsilon R^{n-1}$ is a uniformly distributed random sequence within the range (0, 1). If any of the above generated points are better than the producer point, the producer moves to that point; otherwise, the producer shifts its head to a new randomly generated angle. It repeatedly searches for new locations for better food sources by changing its head angle in every searching bout. Scroungers follow the producer by copying their behaviour using a real coded crossover operator to improve themselves. Scroungers also continue to find their own food sources in every searching bout as per their head angle and search directions. The rest of the less efficient group members, the rangers, perform random walks and systematic search strategies to locate resources efficiently [27]. Rangers help to avoid local maxima or local minima in the search procedure. Ranger X_i movement to a new random point is shown in the following Eq. (5), where $l_i = ar_1 l_{max}$ is random distance.

$$X_i^{k+1} = X_i^k + l_i * D_i^k(\phi^{k+1}) \tag{5}$$

5.2 Modelling GSO to Mine Communities in the Evolutionary Phase

In this chapter, the GSO algorithm is chosen for the evolution phase of the CD framework to evolve communities in relationship networks. The GSO algorithm has the following advantages that make it suitable for handling complex networks.

- A very efficient group searching strategy, which makes it suitable for fitting to the problem of community detection.

- The GSO algorithm optimizes solution using a small population by utilizing its very efficient group food searching strategies. It is advantageous for complex relationship networks, as population size has a direct impact on the complexity of any evolutionary CD algorithm.

GSO performs swarm-based computations to optimize a given population using real-time operators that need to be discretized for its application to discrete optimization problems. As CD is a discrete optimization problem, we need to map (modify) real-time evolution operators to discrete operators in the basic GSO algorithm. Basic GSO modelling to detect communities is discussed below and the modelled GSO algorithm is called the Modified-GSO (Mod-GSO) [8]. Mod-GSO optimizes the population based on the quality function of modularity.

Modified-GSO (Mod-GSO): The GSO algorithm is executed in the evolution phase of the framework to evolve the population initialized using LBA representation for a given complex relationship network. Head angles of all the individuals are initialized, as described by GSO. Size of the population is kept small. Initial population is effectively evolved near to optimal community structure using a modified GSO approach. The main modification applied in Mod-GSO is the conversion of the real coded crossover of scroungers to a single-point discrete crossover. As values in LBA vectors represent node values, scroungers copy producers' vectors without modifying the node values. The Mod-GSO pseudocode and phases are given in Fig. 6.

Input :num: population size; G(v, e): Relationship network of v nodes and e edges; gen = generation
Initialization :
1) Initialize population with individuals X_i: $1 \leq i \leq num$ and $\forall i$: $length (Xi) = v$ using LBA representation.
2) Initialize head angles ϕ_i of all members; initialize count=0.
3) Form clusters for all individuals and calculate their modularity $f(X_i)$: $1 \leq i \leq num$
4) Initialize search variables in GSO manner
Processing:
REPEAT
Selecting :
    Find the producer point X_{prod} with the best fitness: $X_{prod} = max_{1 \leq i \leq n}(f(X_i))$.
    Randomly select 80% of remaining population as scroungers and appoint remaining members as rangers.
Producing:
    X_{prod} optically searches three new points X $_{new1}$, X $_{new2}$, X $_{new3}$at 0^0, left hand and right hand hypercube respectively using equation (2-(4).
    If the fitness of any of the newly generated points is greater than X_{prod}, producer shifts to that point else stays in its own position by shifting its head angle for next iteration.
Scrounging:
    For each scrounger $X_{scroung}$
        Execute Single point crossover to generate two child vectors : X $_{child1}$, X $_{child2}$ =$X_{prod} \times X_{scroung}$
        Turn head angle randomly to search better location in next iteration.
Ranging:
    For each ranger
        Generate a new location by scanning with a random head angle and random distance using equation (5).
Combining:
    Combine all the members and select best 'num' members to generate next generation of population.
    Form clusters for all individuals and calculate their modularity fitness value $f(X_i)$: $1 \leq i \leq num$
Set $count := count + 1$
UNTIL count<gen
Output: Last Generation's Producer's communities

Fig. 6 Pseudocode: Mod-GSO algorithm

5.2.1 Mod-GSO Phases

1. *Selecting*: The solution vector with the best modularity value (best community representation) is selected as a producer. The rest of the population is divided into scroungers and rangers in the ratio of 80:20.
2. *Producing*: The producer scans three new points as described for GSO using Eqs. (2)–(4), while detecting their boundaries using a mod operator to obtain values within permissible node values for a chosen index. If any of the generated points has a better fitness value than the producer, it either shifts to the newly generated point or remains without a change in its vector.
3. *Scrounging and Ranging*: Mod-GSO modifies the area-copying behaviour of scroungers in GSO. GSO uses the real point crossover mechanism to copy the area of the producers' position vector, whereas Mod-GSO uses a single point crossover of standard GA, as shown in Fig. 4 above. A single point crossover helps to copy the position vector values of producer to scroungers, making them fitter and increasing their probability of being selected in the next iteration. Apart from area-copying, the newly generated scroungers also scan their own conic fields for generation of new, better points. Rangers perform random movement to randomly generated positions, as described for GSO in Eq. (5).
4. *Combining*: All the newly generated individuals are combined and top best 'num' number of individuals are then selected to form next generation.

5.3   Experimental Evaluation of Mod-GSO

Parameters values: Population = 48, Number of iterations = 400 and the results are averaged over 20 runs.

Initialization: Head angle a, maximum turning angle α_{max}, maximum pursuit angle θ_{max}, maximum pursuit distance l_{max} and head angle $= n/4$ for all $n - 1$ dimension of an individual's search space are fixed as defined for GSO in [8, 27].

Datasets used: Mod-GSO is tested on real-world and synthetic relationship networks. Real-world networks are: Zachary's Karate Club network (34 nodes, 78 edges) [46], the Bottlenose Dolphin network (62 nodes, 159 edges) [32], the American College Football network (115 nodes, 613 edges) [23], and the Political Books network (105 nodes, 441 edges) [30]. Eleven synthetic variations of classic GN networks proposed by Girvan and Newman [23] are generated using LFR networks [31]. Each network contains 128 nodes divided into 4 communities with $1 - \mu$ percent (μ is the mixing parameter) of its total links with in community and μ percent outside its community. The 11 networks generated differ with regard to the value of the mixing parameter μ.

Comparative Results: Modularity results generated by applying Mod-GSO are compared with the CNM [17], Firefly [4], Rosavall and Bergstrom Infomap (RB) [40], Blondel (Multilevel) [13] and GA-Mod [36] algorithms. Algorithms are compared based on modularity, normalized mutual information (NMI) similarity

Table 1 Best modularity values (Q), NMI and number of communities (NC) results

Method	Dolphins			Polbooks			Karate			Football		
	Q	NMI	NC	Q	NMI	NC	Q	NMI	NC	Q	NMI	NC
Mod-GSO	0.5285	0.5221	5	0.5272	0.5602	3	0.4198	0.5878	4	0.6044	0.8923	10
Blondel	0.5185	0.4418	5	0.5204	0.5121	4	0.4188	0.4899	4	0.6045	0.8903	10
RB	0.5247	0.4721	5	0.5228	0.4934	6	0.4188	0.4899	4	0.6005	0.9241	12
CNM	0.4954	0.5115	4	0.5019	0.5308	4	0.4188	0.4899	4	0.5497	0.6977	6

Table 2 Nature-based CD algorithms modularity comparisons

Datasets	Firefly			GA-Mod			Mod-GSO		
	Best Q	Avg Q	Std-dev	Best Q	Avg Q	Std-dev	Best Q	Avg Q	Std-dev
Karate	0.4185	0.4174	0.1e−2	0.4198	0.4198	0	0.4198	0.4198	0
Footballl	0.6011	0.5974	0.3e−2	0.6046	0.6040	0.1e−2	0.6044	0.6019	0.2e−2
Political books	0.5185	0.5180	0.5e−3	0.5256	0.5251	0.8e−3	0.5272	0.5253	0.8e−3
Dolphin	0.5151	0.5149	0.2e−3	0.5285	0.5270	0.5e−3	0.5285	0.5250	0.4e−2

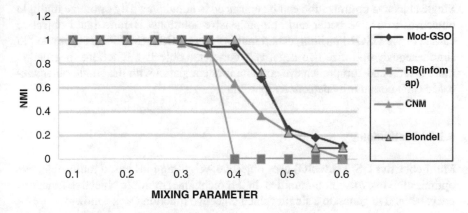

Fig. 7 NMI results on 11 synthetic GN networks

metric [18] and the number of communities (NC) generated (Tables 1 and 2). Actual numbers of communities in the datasets are shown in the NC column heading. Modularities, NMI and NC values generated by Mod-GSO are on the higher side in almost all the data sets, hence showing better community generation by Mod-GSO vis-à-vis other well-known compared algorithms. For synthetic networks, Mod-GSO has shown good results for mixing parameters less than or equal to 3.0 (Fig. 7) as it gives NMI values close to 1. Synthetic networks with low mixing parameter values have well-connected communities, whereas at higher mixing parameter values, these networks tend to have sparse communities. For mixing parameter values 0.3–0.45, Mod-GSO gives higher NMI values than RB and Clauset-Newman-Moore (CNM) and nearly equivalent values to the Blondel algorithm. For higher values of mixing parameters, Mod-GSO shows better NMI results than the RB, CNM and Blondel algorithms (Fig. 7).

6 Variants of GSO

Recent research in the area of nature-based heuristic optimization algorithms has been directed towards studying the effect of varying the base algorithmic behaviours to improve its efficiency in specific application domains. Variations are basically directed and motivated by domain-specific requirements. This section presents some of the dimensions that can be utilized for varying the basic GSO based CD algorithm to achieve better performance in terms of convergence and optimality. Here, four variants of the basic GSO based CD algorithm are presented, which can help to develop intuitions to analyze various possibilities in specific domains to improve the performances of applied algorithms.

6.1 Multi-objective GSO

Single objective optimization can be enhanced to simultaneously optimize multiple objectives to evolve better and comprehensive solutions. Here, a multi-objective variant of the GSO algorithm, called multi-objective GSO (MGSO) is presented to simultaneously optimize two negatively correlated objective functions to generate more natural and comprehensive communities compared with the single objective-based GSO community detection algorithm.

6.1.1 Methodology

Multi-objective GSO (MGSO) uses objective weighting [44] based multi-objective optimization to evolve communities. In this technique, multiple objective functions are combined to generate a single valued objective function O, as follows:

$$O = \sum_{i=1}^{t} w_i f_i(x) \tag{6}$$

Here, x is the set of all possible solutions, f_i is ith objective function such that i ranges from 1 to t, weights w_i are fractional numbers between 0 and 1. To obtain multiple optimal solutions in a single run, various combinations of weight vector are used that provide emphasis to both the optimization functions in varying ratios. Mathematically, a solution obtained with equal weights offers the least objective conflict. The MGSO algorithm uses a combination of (0, 1), (1, 0) and (0.5, 0.5) for the weights of two objectives to generate different possible optimal solutions in a single run. It stores all the generated optimal solutions in a separate repository.

Internal density [38] and community score [35] are the two negatively correlated community quality functions [43] used by MGSO. Community score measures the density of sub-matrices (clusters) based on their volume and power mean, as described in [35]. Internal density is the internal edge density of the cluster [38]. Community score should be maximized to evolve accurate communities, whereas internal density should be minimized. To prune the size of the repository to manageable limits, a fuzzy decision-making policy [1] is used by MGSO in which a linear membership for each objective function is calculated as the following equation.

$$i = \begin{cases} 1 & \text{if } F_i \leq F_i^{\min} \\ \frac{(F_i^{\max}-F_i)}{(F_i^{\max}-F_i^{\min})} & \text{if } F_i^{\min} < F_i < F_i^{\max} \\ 0 & \text{if } F_i \geq F_i^{\max} \end{cases} \tag{7}$$

$F_i^{\max}, F_i^{\min}$ are the maximum and minimum values of the ith objective function respectively. For each of the optimal solutions x in the repository, a normalized membership function μ^x is calculated using the following equation:

$$\mu^x = \frac{\sum_{i=1}^{n} \mu_i^x}{\sum_{j=1}^{k} \sum_{i=1}^{n} \mu_i^j} \tag{8}$$

Here, n is the number of objective functions and k is the number of solutions stored in the repository. The best compromise solution is the one with a maximum value of μ^x. Solutions in the repository are arranged according to their membership value to prune the repository. Optimal solutions stored in the repository are evolved by applying GSO. The MGSO algorithm is shown below (Fig. 8).

MGSO Results: MGSO is executed on real-world and synthetically generated data sets of sizes 128, 500s (small communities), 500b (big communities). Four

1.	Start
2.	Generate initial population of num individuals using LBA.
3.	Form clusters for all individuals.
4.	Repeat steps 'a' to 'e' for max number of generations
	a. Calculate objective1 , objective2 for all individuals in population
	b. Using Objective weighing fill repository with optimum individuals.
	c. Calculate fuzzy membership of repository members
	d. If size of repository is > max then prune it using fuzzy membership.
	e. Select producer, scroungers and rangers based on membership value.
	f. Execute producing, scrounging and ranging as Mod-GSO
	g. Merge the new members to form next generation
5.	Report producer's communities as output.
6.	End

Fig. 8 Pseudocode: the multi-objective GSO (MGSO) algorithm

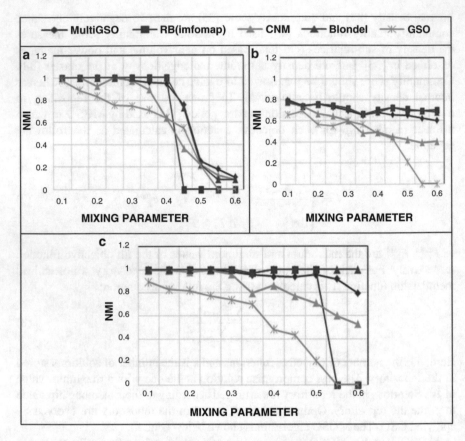

Fig. 9 (**a–c**) NMI values of generated communities in synthetic datasets

real-world datasets are the same as those used for Mod-GSO and two new real-world datasets used here are: Adjnoun [34] and Lesmis [29] (Table 3). MGSO has resulted in more comprehensive and real communities compared with the original GSO and Mod-GSO in almost all the real-world and synthetic datasets considered. NMI comparisons are shown in Fig. 9a (128 dataset), Fig 9b (500s dataset) and Fig 9c (500b dataset). Hence, by optimizing multiple objectives MGSO is able to show improvement over the basic GSO versions of CD algorithms. The following algorithms are compared: CNM [17], multilevel [13] and RB [40].

6.2   TL-GSO

Hybridization of multiple nature-based algorithms has also been studied recently in varying domains of applications [3, 42]. This section explores this dimension

Table 3 Best modularity (Q) and number of communities (NC) results

Datasets	MGSO		GSO		Mod-GSO		RBl		Blondel		CNM	
	Q	NC	Q	NC	Q	NC	Q	NC	Q	NC	Q	NC
Karate	0.4198	4	0.4188	4	0.4198	4	0.4020	3	0.4188	4	0.3806	3
Football	0.6004	12	0.4536	7	0.6044	10	0.6005	12	0.6045	10	0.5497	6
Dolphin	0.5285	5	0.4804	4	0.5285	5	0.5247	5	0.5185	5	0.4954	4
Political books	0.5228	6	0.5149	4	0.5272	5	0.5228	6	0.5204	4	0.5019	4
Lesmis	0.5511	8	0.5449	7	0.5600	6	0.5513	8	0.5555	6	0.5005	5
Adjnoun	0.2989	7	0.2025	10	0.2886	7	0.0092	2	0.2886	7	0.2946	7

for a GSO based CD algorithm by hybridizing it with a new Improved Teachers Learners-Based Optimization (I-TLBO) algorithm [39]. A hybridized version of GSO is referred to here as TL-GSO. Details are presented below.

6.2.1 Methodology

The TL-GSO is a hybrid of GSO and I-TLBO. The hybridization has resulted in fast convergence to accurate communities compared with the results of applying GSO with a single producer. It hybridizes two algorithms by incorporating the exploration capabilities of I-TLBO in the searching strategy of GSO for fast convergence to optimal communities compared with the GSO and Mod-GSO algorithms. I-TLBO [39] is a swarm-based optimization technique that simulates classroom-based teaching and learning methodologies. A percentage of the best fit members are chosen as teachers and the remaining members as students. Students are divided into multiple small groups with individual teachers allocated. Students learn from their respective group teacher and copy teachers' knowledge to improve their own knowledge fitness. Multiple students and teachers resulted in improving the overall fitness of students at a faster rate than with one teacher per student population. TL-GSO uses the same searching principal of GSO producer scrounger (PS) while employing more producers per generation instead of only one. An increase in the number of producers has helped with the fast convergence of the overall best solution compared with the single producer model. Scroungers are divided into groups as per fitness such that the fittest scrounger group is allocated the least fit producer, whereas rangers' functions remained the same as for Mod-GSO. Figure 10 shows the pseudocode of two main phases, i.e. producing and scrounging, of TL-GSO. The rest of the steps are same as those for Mod-GSO.

Results: Increasing the number of producers has resulted in better convergence while maintaining the modularity quality of the communities. The TL-GSO algorithm converges to optimal communities at 200 iterations, which is better than Mod-GSO, which achieved the same modularity results at 400 iterations. Other algorithms compared are Moga-Net [36], Ga-Mod [37], Meme-Net [24], CNM [17], RB [40] and Multilevel [13] (Table 4).

```
Producing :
    1. Select top 5 fittest members from the population as producers.
    2. Producers do Mod-GSO producing
Scrounging :
    1.   Select remaining 75 percent of the best fit members from the remaining
         population as  scroungers.
    2.   Arrange selected scroungers as per their increasing fitness
    3.   Allocate least fit scrounger group to best fit producer and so on.
    4.   Perform single point crossovers of scroungers with their allocated producers.
```

Fig. 10 Pseudocode: TL-GSO producing and scrounging

Table 4 Modularity results
(iterations of GSO-based
algorithms = 200)

Method	Best modularity (Q)			
	Karate	Polbooks	Dolphin	Football
TL-GSO	0.4198	0.5272	0.5285	0.6045
GSO	0.3845	0.4447	0.4298	0.4276
Mod-GSO	0.4198	0.5255	0.5277	0.5900
Ga-Mod	0.4198	0.5256	0.5285	0.6046
Meme-Net	0.4020	0.5232	0.5185	0.6044
Moga-Net	0.4159	0.4993	0.5034	0.4325
CNM	0.3806	0.5019	0.4954	0.5497
Multilevel	0.4188	0.5204	0.5185	0.6045
RB	0.4020	0.5228	0.5247	0.6005

6.3   Discrete TL-GSO

Incorporating discrete behaviour in the evolutionary process can further lead to
improvization in the overall efficiency of the applied algorithm. Here, a discrete
variation of TL-GSO named DTL-GSO. DTL-GSO has shown faster convergence
of the optimization function as compared to the existing variants of GSO and TL-
GSO. It modifies the Mod-GSO based optical search of TL-GSO to step search so as
to discretize the whole optimization process. Thus, producers do discrete step search
producing instead of Mod-GSO producing, as shown in Fig. 10 for TL-GSO. The
modifications result in minimizing the parameters to be externally set. A discrete
step search operator executed by DTL-GSO producers is shown in Fig. 11 for a
random individual of size 8. In an optical search for Mod-GSO, random changes
in the position of the producer vector followed by boundary detection are made,
whereas in discrete step search two positions are randomly selected to create three
partitions of the vector. Elements stored in the three parts of the position vector
are varied in three steps to generate three different position vectors. Variations on a
particular index are made by selecting a random node from the neighbourhood list
of the index nodes under consideration.

Discretization of optical search to step search has led to the improvement
of the convergence speed of TL-GSO. DTL-GSO achieved the near to optimal

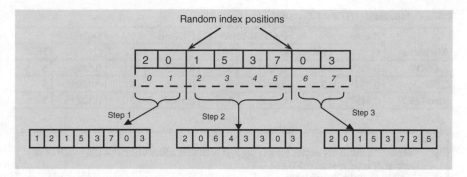

Fig. 11 Three-step discrete search in DTL-GSO

Table 5 Modularity/number of communities detected in the best cases of DTL-GSO and its variants

Method	Karate		Dolphin		Polbooks		Football	
	Q	NC	Q	NC	Q	NC	Q	NC
DTL-GSO	0.4198	4	0.5285	5	0.5272	5	0.6045	10
GSO	0.3953	3	0.4527	5	0.4704	4	0.4123	8
TL-GSO	0.4198	4	0.5124	5	0.5122	5	0.5900	10
Mod-GSO	0.4172	4	0.5086	4	0.4917	4	0.5430	8

modularities in various datasets at 100 generation only, whereas TL-GSO achieved the results at 200 generations and Mod-GSO achieved in 400 iterations (Table 5).

6.4 Enhanced GSO (E-GSO)

Another variant of GSO i.e. E-GSO is presented here, which enhances GSO by merging node similarities in its basic optimization process to fix co-occurrences of highly similar nodes. This novel enhancement has resulted in avoidance of variations on highly similar fixed node positions, enabling faster convergence to optimal communities. The Zhou-Lü-Zhang (ZLZ) node similarity metric [48] is used to calculate node similarities (Eq. (9)). This metric considers two nodes to be similar if they share a greater number of common neighbours. For any pair of connected nodes i and j in a network G, the ZLZ similarity index value Z_{ij} is calculated by summating the inverse of the degrees of all common neighbouring nodes of i and j.

$$Z_{ij} = \begin{cases} \sum_{X \epsilon Nei(i) \cap Nei(j)} \frac{1}{K(X)} & \text{if } i,j \text{ are connected} \\ 0 & \text{otherwise} \end{cases} \quad (9)$$

Table 6 Modularity (Q) and NC results

Algorithm	Karate		Political books		Football		Dolphin	
	Q	NC	Q	NC	Q	NC	Q	NC
E-GSO	0.4197	4	0.5271	5	0.6045	10	0.5285	5
GSO	0.3845	3	0.4447	6	0.4276	7	0.4298	5
Mod-GSO	0.4197	4	0.5255	4	0.5900	7	0.5277	5

```
Start
1.    Generate a similarity vector for storing highly similar nodes of network G which have
      similarity level >= threshold ε.
2.    Initializepopulation of size p
            a)    Create p number of empty individual vectors of size n each: every iᵗʰ
                  position in a vector represents a node in network N.
            b)    Repeat for each iᵗʰ position in the all P individual vectors:
                    •    If the iᵗʰ node position has value in similarity vector then assign
                         it to iᵗʰ position in the position vector V: V[i]=Similarity
                         Vector[i].
                    •    If the node position has no entry in similarity vector then use
                         LBA strategy to assign value to V[i]
            c)    Form clusters for all p individuals and calculate fitness.
Step 3.    Evolve LBA positions of vectors in population using Mod-GSO for max number
           of generations:
Step 4.    Report the clusters of best fit individual at max generation as final best
           communities to the output.
Stop
```

Fig. 12 Pseudocode: E-GSO algorithm

The E-GSO algorithm (Fig. 12) initially calculates node similarity values for all the nodes and stores them in a similarity vector. A threshold is used to fix the level of acceptable similarity value for nodes to fix the attachment in a similarity vector. This similarity vector is then used in the evolution phase of the E-GSO to evolve only those node positions that do not have a connection with a highly similar node having similarity greater than some fixed threshold. The value of the similarity threshold is taken as 0.6 here, as for higher and lower threshold values, a substantial increase in convergence rate was observed. The E-GSO algorithm also achieved near to optimal modularities (Table 6) in varied data sets at 200 iterations, which is an improvement over the GSO and Mod-GSO convergence rates. The E-GSO convergence point matches TL-GSO convergence; hence, we can experiment intuitively with variations in the basic optimization considering different directions.

7 Analyzing Communities: Finding Community Leaders

Communities, once extracted, represent highly interactive functional sub-units within an organization. As organizational activities are group-based, identifying leaders in these functional subgroups is an important task in coordinating the tasks locally within the subgroups and globally across subgroups. Identification of the most active nodes in the overall social network is primarily based on centrality-based quantifiers obtained using social network analytics software. Here, a new novel quantitative measure is proposed to find out community-specific central nodes by considering community-specific centrality in addition to overall influences of nodes. The communities can then be assigned group-specific assignments with identified group representatives (central nodes) as coordinators in their respective groups locally and globally.

7.1 Quantifying Node Importance

Node degree centrality plays an important role in the identification of most connected nodes. Highly connected nodes are more important and help in influencing the neighbourhood. As degree centrality only measures the strength of immediate connections, considering that more influential parameters can benefit from identification of influential nodes that have a wide spread of influences in the cluster. Other important measures of importance worth considering are the nodes betweenness [21] and closeness [12]. Betweenness indicates the importance of a node based on the number of shortest paths on which it appears. Betweenness captures the broker nodes in any network, which can have a small number of immediate connections that are important because of their participating nature by acting as a connector among different clusters of a network and hence providing access to new ideas, opinions and opportunities in a network. The betweenness centrality of a node i is calculated as:

$$B(i) = \sum_{p<q} p_{pq}(i)) p_{pq} \tag{10}$$

Here, p_{pq} is the number of shortest paths connecting any two node pairs p and q and $p_{pq}(i)$ is the number of shortest paths the node i is on. Closeness of a node emphasizes how far the rest of network is from a particular node. Closeness is based on the length of the average shortest path between a node and all other nodes in the network such that $d(p, q)$ is the distance between node p and node q.

$$C(i) = \left[\sum_{j=1}^{N} d(i,j) \right]^{-1} \tag{11}$$

Another important metric for quantifying the importance of a node is by quantifying its authority and hub value using the HITS(hyperlink-induced topic search) algorithm [28]. The authority value signifies the importance of information stored in the node, whereas the hub value signifies the importance of its links. Here, we present a novel metric for extracting community-based central nodes by quantifying node ranks using intra-community degrees and community edge density. As counting only immediate connections can mislead the selection of real influential nodes, the proposed rank calculation uses a network influence value of node as a scaling factor. The network influence of a node is based on its overall closeness, betweenness and hub values, which signifies the influence of a node in an overall network. The metric that calculates the rank of a node i is shown below.

$$\text{node\_rank}_i = \frac{\text{community\_degree}(i)}{\text{community\_density}} * \text{Inf}_i \tag{12}$$

$$\text{Inf}_i = (\text{avg}(B(i), C(i)) * (\text{hub}(i)) + \beta) \tag{13}$$

In the above equation community_degree(i) is the number of connections that node i has in its community, community_density is the total number of edges within the node i community, network_influence$_i$ is the overall influence of the node i and is calculated as shown in Eq. (13). $B(i)$ is the betweenness of node i, $C(i)$ is the closeness of node i and hub (i) is hub value of node i. Here, β is the additive constant set to 1 that ensures a minimum hub value. We present here results using the Karate Club dataset, for which the GSO-based CD algorithm has detected four communities. Betweenness, closeness and hub values of all the nodes are imported from Gephi software. Distributions of nodes in GSO communities at 0.4197 modularity are as follows:

community 1 = [0, 1, 2, 3, 7, 11, 12, 13, 17, 19, 21]
community 2 = [16, 10, 4, 5, 6]
community 3 = [32, 33, 8, 9, 14, 15, 18, 20, 22, 26, 29, 30]
community 4 = [23, 24, 25, 27, 28, 31]

Node influences calculated by using betweenness, closeness and hub metric values imported from Gephi software for the Karate Club dataset are shown below.

Node influences = 0: 1.0, 1: 1.5, 2: 5.5, 3: 2.5, 4: 1.5, 5: 1.5, 6: 2.0, 7: 1.5, 8: 2.5, 9: 1.5, 10: 1.5, 11: 1.5, 12: 1.5, 13: 2.0, 14: 1.0, 15: 1.0, 16: 1.5, 17: 1.5, 18: 1.0, 19: 1.5, 20: 1.0, 21: 1.5, 22: 1.0, 23: 1.0, 24: 1.0, 25: 2.0, 26: 1.0, 27: 1.5, 28: 2.5, 29: 2.0, 30: 1.5, 31: 4.0, 32: 1.5, 33: 1.5

Table 7 shows the node rank values calculated for each node in column R, the total degree column shows the total number of edges whereas in-degree column shows the number of edges that a node shares with-in its community.

Table 7 Node rank table: Karate Club relationship network

Community	Node	Total degree	In Degree	R	Community	Node	Total degree	In degree	R
1	0	16	10	0.20	3	32	12	9	0.321
	1	9	8	0.25		33 (Leader)	17	11	0.392
	2 (Leader)	10	5	0.597		8	5	3	0.178
	3	6	6	0.326		14	2	2	0.04
	7	4	4	0.130		15	2	2	0.04
	11	1	1	0.032		18	2	2	0.04
	12	2	2	0.065		20	2	2	0.04
	13	5	4	0.173		22	2	2	0.04
	17	2	2	0.065		26	2	2	0.04
	19	3	2	0.065		29	4	3	0.14
	21	2	2	0.065		30	4	3	0.10
2	5	4	3	0.375		9	2	1	0.03
	4	3	2	0.25	4	23	5	2	0.142
	10	3	2	0.25		24	3	3	0.214
	6 (Leader)	4	3	0.5		25	3	3	0.428
	16	2	2	0.25		27	4	2	0.214
						28	3	1	0.178
						31(Leader)	6	3	0.857

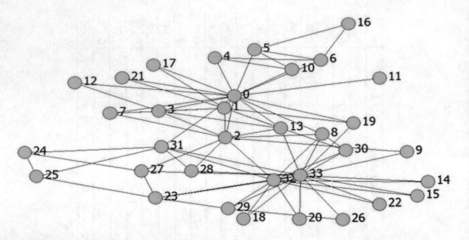

Fig. 13 Karate Club relationship network

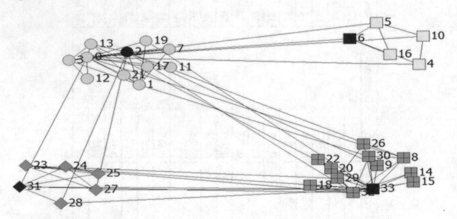

Fig. 14 Karate Club GSO extracted communities with leaders

The Table 7 shows that, even if node 2 was low on within-community degree, its overall influence made it an important node to take the role of the community group leader. Such a selection would enable the leaders to have interactions and influences on other groups, leading to improving the local performance of the community to have global improvement. In community 2, node 6 and node 5 have the same number of indegrees, but the overall influence of node 6 is higher, hence making it leader of community 2. Figure 13 shows the actual Karate Club relationships and Fig. 14 shows its community partitions having evolved using the DTL-GSO algorithm and their respective identified community leaders. Netdraw software [14] is used to plot the evolved communities. Similarly, leaders for communities 3 and 4 are chosen based on the higher ranks. Similar experiments were conducted on the other real-world and artificial networks mentioned in earlier sections and community leaders were successfully assigned.

8 Case Study: School Children Social Network

Here, a real-life case study using a school children's relationship network dataset is presented. The dataset created by Fournet and Barrat [20] gives the contacts of the students of three classes in a high school in Marseilles, France, over 4 days in December 2011 for 126 students. Electronic body sensor technology was used by attaching them to students to record the number and time of contacts among the children of the three classes. Contact lists were created according to the data captured by the sensors. Contacts were only recorded for interactions lasting for a minimum of 20 s. The contact list file contained a tab-separated list representing the active contacts during 20-s intervals of the data collection. In the contacts file, each contact has the following information: t i j Ci Cj, where i and j are the IDs of the students in contact, Ci and Cj are their classes, and the interval during which this contact was active is [t-20s, t]. Time is measured in seconds. These recorded contact data are converted into graph format by adding nodes and edges for all contacts to create a network of 126 nodes and 1710 valid edges using Python 2.7. The contact network generated was then evolved using GSO based community detection and other CD algorithms for comparative analysis. Community-specific leaders are also identified by applying the proposed leader identification rank metric for group-specific assignment of tasks or projects. Figure 15 shows the initial network generated from the data file of the students' contacts.

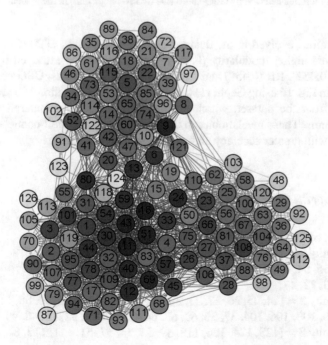

Fig. 15 Actual student interaction network

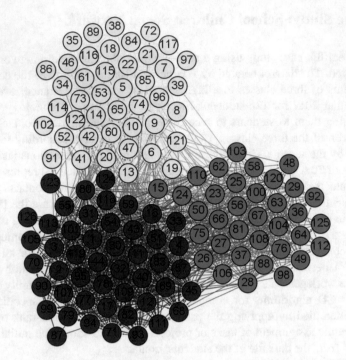

Fig. 16 Communities detected by GSO-based (DTL-GSO) algorithm in the students' network

Communities evolved in the data set using a GSO-based DTL-GSO CD algorithm showed higher modularity (0.459) than communities extracted by applying the CNM (0.452), RB (0.458) and Blondel (0.458) algorithms. GSO communities are shown in Fig. 16 using Gephi [11]. This signifies the extraction of better modular structures from the dataset, which are densely connected compared with other CD algorithms. These revelations can help to identify real, well-connected student subgroups with greater accuracy and assign group-specific tasks.

8.1   Experimental Results

- GSO based DTL-GSO: Three Communities at modularity 0.459
  community 1 = [20, 21, 22, 8, 121, 122, 53, 52, 115, 114, 116, 86, 84, 85, 7, 102, 39, 38, 35, 34, 60, 61, 65, 6, 91, 97, 96, 10, 13, 14, 19, 18, 117, 89, 46, 42, 41, 5, 9, 74, 73, 72, 47]
  community 2 = [24, 25, 26, 27, 23, 28, 29, 120, 125, 58, 56, 50, 88, 110, 112, 81, 108, 103, 100, 106, 104, 37, 36, 62, 63, 64, 66, 67, 98, 92, 15, 48, 49, 76, 75]
  community 3 = [123, 124, 126, 118, 59, 55, 54, 57, 51, 111, 113, 82, 83, 80, 87,

3, 109, 101, 107, 105, 33, 32, 31, 30, 68, 69, 2, 99, 90, 93, 95, 94, 11, 12, 17, 16, 119, 44, 45, 43, 40, 1, 77, 71, 70, 79, 78]
- Multilevel (Blondel): Three Communities at modularity = 0.4586
  community 1 = [4, 5, 6, 11, 13, 14, 28, 29, 30, 31, 33, 42, 44, 45, 47, 50, 58, 59, 66, 67, 68, 69, 73, 79, 82, 88, 89, 91, 92, 95, 98, 99, 100, 101, 104, 108, 111, 113, 114, 118, 119, 120, 125]
  community 2 = [0, 1, 2, 3, 7, 8, 9, 12, 17, 21, 25, 27, 32, 35, 37, 41, 48, 51, 52, 54, 56, 64, 65, 70, 71, 72, 74, 75, 81, 85, 94, 102, 103, 116, 117]
  community 3 = [10, 15, 16, 18, 19, 20, 22, 23, 24, 26, 34, 36, 38, 39, 40, 43, 46, 49, 53, 55, 57, 60, 61, 62, 63, 76, 77, 78, 80, 83, 84, 86, 87, 90, 93, 96, 97, 105, 106, 107, 109, 110, 112, 115, 121, 122, 123, 124]
- CNM: Three communities at modularity = 0.4529
  community 1 = [0, 1, 2, 3, 7, 8, 9, 10, 12, 17, 21, 25, 27, 32, 35, 37, 41, 48, 51, 52, 54, 56, 60, 64, 65, 70, 71, 72, 74, 75, 81, 85, 94, 102, 103, 116, 117]
  community 2 = [4, 5, 6, 11, 13, 14, 28, 29, 30, 31, 33, 42, 44, 45, 47, 50, 58, 59, 66, 67, 68, 69, 73, 79, 82, 88, 89, 91, 92, 95, 98, 99, 100, 101, 104, 108, 111, 113, 114, 118, 119, 120, 125]
  community 3 = [15, 16, 18, 19, 20, 22, 23, 24, 26, 34, 36, 38, 39, 40, 43, 46, 49, 53, 55, 57, 61, 62, 63, 76, 77, 78, 80, 83, 84, 86, 87, 90, 93, 96, 97, 105, 106, 107, 109, 110, 112, 115, 121, 122, 123, 124]
- Infomap (Rosavall and Bergstrom): Three communities at modularity = 0.4586
  community 1 = [10, 15, 16, 18, 19, 20, 22, 23, 24, 26, 34, 36, 38, 39, 40, 43, 46, 49, 53, 55, 57, 60, 61, 62, 63, 76, 77, 78, 80, 83, 84, 86, 87, 90, 93, 96, 97, 105, 106, 107, 109, 110, 112, 115, 121, 122, 123, 124]
  community 2 = [4, 5, 6, 11, 13, 14, 28, 29, 30, 31, 33, 42, 44, 45, 47, 50, 58, 59, 66, 67, 68, 69, 73, 79, 82, 88, 89, 91, 92, 95, 98, 99, 100, 101, 104, 108, 111, 113, 114, 118, 119, 120, 125]
  community 3 = [0, 1, 2, 3, 7, 8, 9, 12, 17, 21, 25, 27, 32, 35, 37, 41, 48, 51, 52, 54, 56, 64, 65, 70, 71, 72, 74, 75, 81, 85, 94, 102, 103, 116, 117]

8.1.1 Group Leaders Extracted for GSO Communities: Results

Community-specific group leaders are also extracted in students' communities to enable the task of project assignment. The node rank metric discussed in the above section is applied to all the nodes to identify community-specific central nodes. Tables 8 and 9 shows the identified community leaders highlighted i.e. nodes 53, 106 and 93 as leaders for communities 1, 2 and 3 respectively. Actual and within community node degrees of the identified leader nodes are also highlighted in Tables 8 and 9 to emphasize the fact that degree is not the only criterion for designating a specific node as central, but its overall influence plays a significant role in defining centrality.

Table 8 Actual node degrees: students contact network

Community 1	Community 2	Community 3
Actual node degrees		
[20 - 45, 21 - 25, 22 - 29, 8 - 32, 121 - 29, 122 - 15, 53 - 42, 52 - 26, 115 - 36, 114 - 50, 116 - 22, 86 - 34, 84 - 21, 85 - 11, 7 - 39, 102 - 7, 39 - 26, 38 - 26, 35 - 27, 34 - 18, 60 - 42, 61 - 31, 65 - 23, 6 - 34, 91 - 15, 97 - 55, 96 - 27, 10 - 29, 13 - 27, 14 - 7, 19 - 12, 18 - 4, 117 - 32, 89 - 15, 46 - 40, 42 - 10, 41 - 31, 5 - 19, 9 - 22, 74 - 40, 73 - 31, 72 - 21, 47 - 15]	[24 - 44, 25 - 26, 26 - 51, 27 - 21, 23 - 43, 28 - 22, 29 - 36, 120 - 10, 125 - 27, 58 - 21, 56 - 28, 50 - 5, 88 - 6, 110 - 47, 112 - 40, 81 - 9, 108 - 33, 103 - 20, 100 - 18, 106 - 44, 104 - 20, 37 - 8, 36 - 15, 62 - 34, 63 - 44, 64 - 35, 66 - 19, 67 - 26, 98 - 20, 92 - 46, 15 - 4, 48 - 40, 49 - 34,76 - 22, 75 - 31]	[123 - 22, 124 - 39, 118 - 31, 59 - 24, 55 - 26, 54 - 34, 57 - 14, 51 - 11, 111 - 28, 113 - 37, 82 - 12, 83 - 27, 80 - 12, 87 - 31, 3 - 34, 109 - 53, 101 - 16, 107 - 41, 105 - 16, 33 - 14, 32 - 26, 31 - 20, 30 - 35, 68 - 32, 69 - 17, 2 - 38, 99 - 24, 90 - 54, 93 - 48, 95 - 27, 94 - 22, 11 - 30, 12 - 15, 17 - 7, 16 - 2, 119 - 31, 44 - 20, 45 - 25, 43 - 4, 40 - 46, 1 - 27, 77 - 50, 71 - 25, 70 - 26, 79 - 44, 78 - 30]

Table 9 Community node degrees: students' contact network

Community 1	Community 2	Community 3
Node degrees with in the community		
[5 - 2, 6 - 8, 7 - 16, 8 - 13, 9 - 8, 10 - 6, 13 - 7, 14 - 2, 18 - 1, 19 - 6, 20 - 18, 21 - 12, 22 - 7, 34 - 5, 35 - 6, 38 - 10, 39 - 7, 41 - 12, 42 - 4, 46 - 13, 47 - 2, 52 - 11, 53 - 16, 60 - 15, 61 - 10, 65 - 11, 72 - 8, 73 - 9, 74 - 13, 84 - 7, 85 - 4, 86 - 11, 89 - 4, 91 - 3, 96 - 12, 97 - 20, 102 - 1, 114 - 12, 115 - 13, 116 - 10, 117 - 11, 121 - 10, 122 - 4]	[15 - 1, 23 - 10, 24 - 14, 25 - 5, 26 - 11, 27 - 4, 28 - 6, 29 - 10, 36 - 4, 37 - 2, 48 - 9, 49 - 10, 50 - 0, 56 - 5, 58 - 7, 62 - 10, 63 - 14, 64 - 8, 66 - 7, 67 - 7, 75 - 8, 76 - 6, 81 - 3, 88 - 1, 92 - 11, 98 - 6, 100 - 4, 103 - 6, 104 - 8, 106 - 10, 108 - 12, 110 - 12, 112 - 9, 120 - 1, 125 - 11]	[1 - 8, 2 - 12, 3 - 9, 11 - 13, 12 - 2, 16 - 2, 17 - 2, 30 - 15, 31 - 8, 32 - 8, 33 - 8, 40 - 19, 43 - 2, 44 - 7, 45 - 10, 51 - 5, 54 - 12, 55 - 12, 57 - 5, 59 - 12, 68 - 15, 69 - 10, 70 - 8, 71 - 5, 77 - 15, 78 - 11, 79 - 18, 80 - 3, 82 - 5, 83 - 10, 87 - 9, 90 - 21, 93 - 15, 94 - 9, 95 - 12, 99 - 9, 101 - 7, 105 - 4, 107 - 12, 109 - 19, 111 - 11, 113 - 17, 118 - 13, 119 - 13, 123 - 9, 124 - 12]

9 Conclusion

Complex networked environments appear as hairballs and hence increase the complexity of intuitively identifying logically operational hidden subunits. It is imperative to identify functional subgroups (communities) along with their leaders to gain a deeper insight into these networks, thereby enabling coordination and distribution of organizational activities efficiently and discretely. This chapter provides a comprehensive discussion on the applicability of nature-based heuristics to mine communities, thereby enabling identification of community-specific leaders. A novel node_rank metric is proposed in this chapter to extract the most influential nodes from the identified communities that can have a great influence on coordinating tasks in and across the identified communities. The proposed metric utilizes local community-wise degree density in addition to global influencing factors, such as closeness, betweenness and hub metrics, to rank community nodes to identify highly trusted leaders in these communities. As the optimality of communities plays an important role in perfectly partitioning these networks, an efficient DTL-GSO community detection algorithm is applied for the detection of optimal communities. DTL-GSO is a recently proposed nature-based optimization algorithm that has been proven to detect highly optimal communities using group food searching behaviours of animals. Many other GSO based heuristics are also discussed in detail using a framework-based approach to understanding directions and possibilities that can be effectively experimented with in other domains. Experimental analysis is provided, using many real-world and artificial network data sets. A case study on a school children's contact relationship network was made to experimentally test community generation and group leader extraction.

References

1. Abilo, M.A.: Multiobjective evolutionary algorithms for electric power dispatch problem. IEEE Trans. Evol. Comput. **10**(3), 315–329 (2006)
2. Ahmed, K., Hafez, A.I., Hassanien, A.E.: A discrete Krill herd optimization algorithm for community detection. In: 2015 11th International Computer Engineering Conference (ICENCO), December, pp. 297–302. IEEE, New York (2015)
3. Amiri, B., Hossain, L., Crawford, J.: A hybrid evolutionary algorithm based on HSA and CLS for multi-objective community detection in complex networks. In: 2012 IEEE/ACM International Conference on Advances In Social Networks Analysis and Mining (ASONAM), August, pp. 243–247. IEEE, New York (2012)
4. Amiri, B., Hossain, L., Crawford, J.W., Wigand, R.T.: Community detection in complex networks: multiobjective enhanced firefly algorithm. Knowl.-Based Syst. **46**, 1–11 (2013)
5. Arora, N., Banati, H.: Multiobjective group search optimization approach for community detection in networks. Int. J. Appl. Evol. Comput. (IJAEC) **7**(3), 50–70 (2016)
6. Arora, N., Banati, H.: Enhancing group search optimization with node similarities for detecting communities. In: The International Symposium on Intelligent Systems Technologies and Applications, pp. 303–316, September. Springer International Publishing, Cham (2016)
7. Banati, H., Arora, N.: Enabling inclusive education in structured learning environments through social network analysis. Int. J. Innov. Educ. **2**(2–4), 151–167 (2014)
8. Banati, H., Arora, N.: Modeling evolutionary group search optimization approach for community detection in social networks. In: Proceedings of the Third International Symposium on Women in Computing and Informatics, August, pp. 109–117. ACM, New York (2015)
9. Banati, H., Arora, N.: TL-GSO: a hybrid approach to mine communities from social networks. In: 2015 IEEE International Conference on Research in Computational Intelligence and Communication Networks (ICRCICN), November, pp. 145–150. IEEE, New York (2015)
10. Banati, H., Arora, N.: Detecting communities in complex networks-A discrete hybrid evolutionary approach. Int. J. Comput. Appl. **38**(1), 29–40 (2016)
11. Bastian, M., Heymann, S., Jacomy, M.: Gephi: an open source software for exploring and manipulating networks. ICWSM **8**, 361–362 (2009)
12. Bavelas, A.: Communication patterns in task-oriented groups. J. Acoust. Soc. Am. **22**(6), 725–730 (1950)
13. Blondel, V.D., Guillaume, J.L., Lambiotte, R., Lefebvre, E.: Fast unfolding of communities in large networks. J. Stat. Mech.: Theor. Exp. **2008**(10), P10008 (2008)

14. Borgatti, S.P.: NetDraw Software for Network Visualization. Analytic Technologies, Lexington, KY (2002)
15. Cao, C., Ni, Q., Zhai, Y.: A novel community detection method based on discrete particle swarm optimization algorithms in complex networks. In: 2015 IEEE Congress on Evolutionary Computation (CEC), May, pp. 171–178. IEEE, New York (2015)
16. Chen, D., Wang, D., Xia, F.: A modularity degree based heuristic community detection algorithm. Math. Probl. Eng. **2014**, 9 pp. (2014)
17. Clauset, A., Newman, M.E., Moore, C.: Finding community structure in very large networks. Phys. Rev. E **70**(6), 066111 (2004)
18. Danon, L., Diaz-Guilera, A., Duch, J., Arenas, A.: Comparing community structure identification. J. Stat. Mech.: Theor. Exp. **2005**(09), P09008 (2005)
19. Dasgupta, D., Michalewicz, Z.: Evolutionary algorithms in engineering applications. Int. J. Evol. Optim. **1**, 93–94 (1999)
20. Fournet, J., Barrat, A.: Contact patterns among high school students. PLoS One **9**(9), e107878 (2014)
21. Freeman, L.C.: A set of measures of centrality based on betweenness. Sociometry **40**(1), 35–41 (1977)
22. Gach, O., Hao, J.K.: A memetic algorithm for community detection in complex networks. In: International Conference on Parallel Problem Solving from Nature, September, pp. 327–336. Springer, Berlin, Heidelberg (2012)
23. Girvan, M., Newman, M.E.: Community structure in social and biological networks. Proc. Natl. Acad. Sci. **99**(12), 7821–7826 (2002)
24. Gong, M., Fu, B., Jiao, L., Du, H.: Memetic algorithm for community detection in networks. Phys. Rev. E **84**(5), 056101 (2011)
25. Gong, M., Cai, Q., Chen, X., Ma, L.: Complex network clustering by multiobjective discrete particle swarm optimization based on decomposition. IEEE Trans. Evol. Comput. **18**(1), 82–97 (2014)
26. Hafez, A.I., Zawbaa, H.M., Hassanien, A.E., Fahmy, A.A.: Networks community detection using Artificial bee colony swarm optimization. In: Proceedings of the Fifth International Conference on Innovations in Bio-Inspired Computing and Applications IBICA 2014, pp. 229–239. Springer International Publishing, Cham (2014)
27. He, S., Wu, Q.H., Saunders, J.R.: Group search optimizer: an optimization algorithm inspired by animal searching behavior. IEEE Trans. Evol. Comput. **13**(5), 973–990 (2009)
28. Kleinberg, J.M.: Authoritative sources in a hyperlinked environment. J. ACM **46**(5), 604–632(1999)
29. Knuth, D.E.: The Stanford GraphBase: A Platform for Combinatorial Computing, vol. 37. Addison-Wesley, Reading (1993)
30. Kreb, V.: (2008). http://www.orgnet.com/cases.html
31. Lancichinetti, A., Fortunato, S., Radicchi, F.: Benchmark graphs for testing community detection algorithms. Phys. Rev. E **78**(4), 046110 (2008)
32. Lusseau, D., Schneider, K., Boisseau, O.J., Haase, P., Slooten, E., Dawson, S.M.: The bottlenose dolphin community of Doubtful Sound features a large proportion of long-lasting associations. Behav. Ecol. Sociobiol. **54**(4), 396–405 (2003)
33. Newman, M.E.: Fast algorithm for detecting community structure in networks. Phys. Rev. E **69**(6), 066133 (2004)
34. Newman, M.E.: Finding community structure in networks using the eigenvectors of matrices. Phys. Rev. E **74**(3), 036104 (2006)
35. Pizzuti, C.: Ga-net: a genetic algorithm for community detection in social networks. In: International Conference on Parallel Problem Solving from Nature, September, pp. 1081–1090. Springer, Berlin, Heidelberg (2008)
36. Pizzuti, C.: A multiobjective genetic algorithm to find communities in complex networks. IEEE Trans. Evol. Comput. **16**(3), 418–430 (2012)

37. Pizzuti, C.: Boosting the detection of modular community structure with genetic algorithms and local search. In: Proceedings of the 27th Annual ACM Symposium on Applied Computing, March, pp. 226–231. ACM, New York (2012)
38. Radicchi, F., Castellano, C., Cecconi, F., Loreto, V., Parisi, D.: Defining and identifying communities in networks. Proc. Natl. Acad. Sci. U. S. A. **101**(9), 2658–2663 (2004)
39. Rao, R.V., Patel, V.: An improved teaching-learning-based optimization algorithm for solving unconstrained optimization problems. Sci. Iran. **20**(3), 710–720 (2013)
40. Rosvall, M., Bergstrom, C.T.: Maps of random walks on complex networks reveal community structure. Proc. Natl. Acad. Sci. **105**(4), 1118–1123 (2008)
41. Schaeffer, S.E.: Graph clustering. Comput. Sci. Rev. **1**(1), 27–64 (2007)
42. Shi, X.H., Lu, Y.H., Zhou, C.G., Lee, H.P., Lin, W.Z., Liang, Y.C.: Hybrid evolutionary algorithms based on PSO and GA. In: The 2003 Congress on Evolutionary Computation, December, vol. 4, pp. 2393–2399. IEEE, New York (2003)
43. Shi, C., Yu, P.S., Yan, Z., Huang, Y., Wang, B.: Comparison and selection of objective functions in multiobjective community detection. Comput. Intell. **30**(3), 562–582 (2014)
44. Srinivas, N., Deb, K.: Multiobjective optimization using non-dominated sorting in genetic algorithms. Evol. Comput. **2**(3), 221–248 (1994)
45. Tapia, M.G.C., Coello, C.A.C.: Applications of multi-objective evolutionary algorithms in economics and finance: a survey. In: IEEE Congress on Evolutionary Computation, September, vol. 7, pp. 532–539 (2007)
46. Zachary, W.W.: An information flow model for conflict and fission in small groups. J. Anthropol. Res. **33**(4), 452–473 (1977)
47. Zhang, C., Hei, X., Yang, D., Wang, L.: A memetic particle swarm optimization algorithm for community detection in complex networks. Int. J. Pattern Recogn. Artif. Intell. **30**(02), 1659003 (2016)
48. Zhou, T., Lü, L., Zhang, Y.C.: Predicting missing links via local information. Eur. Phys. J. B-Cond. Matt. Com. Syst. **71**(4), 623–630 (2009)

A Holistic Approach to Influence Maximization

Nireshwalya Sumith, Basava Annappa, and Swapan Bhattacharya

Abstract A social network is an Internet-based collaboration platform that plays a vital role in information spread, opinion-forming, trend-setting, and keeps everyone connected. Moreover, the popularity of web and social networks has interesting applications including viral marketing, recommendation systems, poll analysis, etc. In these applications, user influence plays an important role. This chapter discusses how effectively social networks can be used for information propagation in the context of viral marketing. Picking the right group of users, hoping they will cause a chain effect of marketing, is the core of viral marketing applications. The strategy used to select the correct group of users is the influence maximization problem.

This chapter proposes one of the viable solutions to influence maximization. The focus is to find those users in the social networks who would adopt and propagate information, thus resulting in an effective marketing strategy. The three main components that would help in the effective spread of information in the social networks are: the network structure, the user's influence on others, and the seeding algorithm. Amalgamation of these three aspects provides a holistic solution to influence maximization.

Keywords Algorithm • Diffusion • Influence maximization • Social networks • Viral marketing

1 Introduction

With the advent of Web 2.0 came a range of applications that are used in many ways by people across different sections of the society. The social network is one such application that plays a very important role across the world. It is not just a platform for sharing ideas, it is also seen to play an important role in the economic growth of

N. Sumith (✉) • B. Annappa
National Institute of Technology Karnataka, Surathkal, Mangaluru, India
e-mail: s.nireshwalya@gmail.com; annappa@ieee.org

S. Bhattacharya
Jadavpur University, Kolkata, India
e-mail: bswapan2000@yahoo.co.in

© Springer International Publishing AG 2017   129
H. Banati et al. (eds.), *Hybrid Intelligence for Social Networks*,
DOI 10.1007/978-3-319-65139-2_6

the country. The term social economics reflects the importance of social networks in economic transactions. In the era of cloud computing, social media has proved to be a more effective business-related strategy [36].

Often, influence among friends plays an important role in product adoption decisions. An individual's choice to adopt or reject a product is often linked to his/her peers' choices. The term network externalities embodies such choices. Undoubtedly, this trend is seen in social networks and is used to popularize a product in the network to increase sales. An early attempt to model network structures, with some perspective on their impact on economic outcomes, is seen in the cooperative game theory literature. The game theory relies on the premise that users can cooperate only when they are connected. To understand these connections, graphs are employed. People who can communicate can cooperate and generally cooperation leads to higher production or utility than separate efforts [37]. Thus, graph representations became an important part of game theory and social network analysis.

In this chapter, the role of social networks in the context of viral marketing is discussed. The success of viral marketing depends on the strategy used to select initial adopters, network structure, and Influence among users. These aspects are discussed in detail in this chapter. An outline of the chapter is as follows. Section 2, discusses various cases in which viral marketing is employed. Section 3 introduces the influence maximization problem. Section 4 analyzes social networks to obtain a new evaluation metric, followed by details of the proposed approach in Sect. 5. Section 6 summarizes results and a conclusion constitutes Sect. 7.

2 Viral Marketing in the Real World

With billions of users of social network sites, they have become the most powerful tool for marketing. User involvement has made viral marketing, tailored to social networks, to be more dominant than the traditional marketing approaches. The strategy where individuals forward the message to others, creating a vast spread of information and influencing others to adopt it and spread further is popularly referred to as viral marketing. The brand awareness thus created by viral marketing is cost-effective and generates requests for products. The practice of viral marketing in the digital era has been around for more than a decade. The early adopters of the viral marketing strategy were Hotmail, which grew to 12 million users in 18 months, and the John West salmon bear advertisement, to name a few. Although these campaigns were accidentally successful, they were not well planned. The low expenditure on popularizing products is the main reason for enterprises to adopt this strategy. In the following sections, three popular cases across various domains, in which viral marketing created a success story are discussed.

2.1   Case Study 1: Fiesta Ford Movement

Ford had made several attempts to market a small car since the discontinuation of the Aspire in 1997, but without much success. In 2009, Ford Motors launched the Fiesta Movement campaign [34] to promote sales. For 6 months, Ford gave 100 people a car to use and asked them to write about their experiences on social media. Consumers used their Fiestas for various activities and some went for adventures. These consumers wrote about their experiences on YouTube, Flickr, Facebook, and Twitter. The social media audience took great interest in these blogs and soon it resulted in massive sales of the Fiesta.

The Fiesta Movement was the most successful social media marketing experiment for the automotive world. The campaign news was all over the social media with 6.5 million YouTube views alone and 50,000 queries about the car from new customers. In first week of the campaign, Ford sold nearly 10,000 cars. The Fiesta Movement cost the company only a small amount compared with the typical traditional TV campaign. In 2014, Ford used this strategy to introduce their latest Fiesta.

2.2   Case Study 2: Why So Serious?

In 2008, the "Why So Serious?" campaign, an augmented reality game (ARG) was launched to promote the movie, *The Dark Knight* [52]. Over 10 million people participated in this campaign, which was launched 15 months before the release of the movie. Various games and rewards were available all over social media and participants took great interest in these. The ARG was thus able to maintain fan interest up to the release of the movie. Millions of blog posts were seen on social media, resulting in success of the ARG and leading to the success of the film, which made over US$ 1 billion in box office collections.

The Dark Knight Rises promotion also used a similar campaign. This time the participants were given graffiti to help the Gotham City Police Department find *Batman*. For every piece of graffiti found and tagged on social media, a frame of the trailer would be released. This marketing strategy, because of the massive fan interest, led to completion of the task within a few hours.

2.3   Case Study 3: Ice Bucket Challenge

In 2014, to promote awareness of Amyotrophic Lateral Sclerosis (ALS), the "The Ice Bucket Challenge" campaign was designed [14]. In this challenge, a person needed to pour a bucket of iced water over their head, film it and upload it. A person who did not accept the challenge had to donate to ALS cause within 24 h. Once the participant had either been soaked or had donated, this challenge had to be passed to three friends.

This campaign was popular on Facebook and Twitter, with over 2.4 million tagged videos and 2.2 million Tweets respectively, about the challenge. The views per month grew from 0.16 million views, to over 2.89 million per month in August 2014. Because of this, huge donations to ALS were received. The ALS fund had received over $40 million from seven hundred thousand donors within 30 days. The ALS association declared the total donation received to be around $100 million.

There are a number of similar successful cases where a social network was used to effectively promote information for various causes. User involvement in social networks is the driving force behind these successful campaigns. In the following sections, viral marketing is presented as an optimization problem and a solution is proposed.

3 Influence Maximization Problem

There are many cases in which enterprises have created a success story with a viral marketing strategy. The keys to these viral marketing campaigns are those first few users who started the campaign. These initial users were picked by the enterprise based on various criteria. Whatever the strategy was, the outcome was aimed at creating successful results. Therefore, these individuals should be picked with proper planning. Picking these individuals is referred to as the influence maximization problem. Figure 1 shows the process of information propagation by the initial user. Influence maximization is aimed to obtaining a good-quality seed set to maximize the spread of information in the social network. Formally, the problem discussed in this chapter is defined as follows.

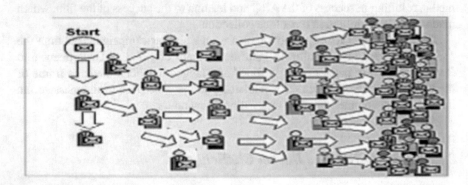

Fig. 1 Information spread phenomenon in social network

Influence maximization [57]: Given a cost k and a social network, which is represented as a directed graph $G = (V, E)$, the goal is to find a seed set of k users such that by initially targeting them, the expected influence spread (in terms of expected number of adopted users) can be maximized.

Social networks play a fundamental role as a medium for the diffusion of information and ideas. This diffusion of information can be modelled to understand and answer many of the questions that arise in the real-world application. The independent cascade model (ICM) is a popular model used to understand the diffusion process in the network and is explained as follows.

Independent Cascade Model[24]: Suppose that node u is influenced (i.e., becomes active) at a time t. Then, u has an opportunity to influence every one of its neighbors v with probability $p(u,v)$. If u succeeds in activating v, then v is active from time $t+1$ onwards. If not, u can never try to influence v in subsequent attempts. This process continues until no new node becomes active at the end of the diffusion process.

4 Analyzing the Social Network

The initial part of the section discusses various existing approaches to evaluating a user to rank him/her to be the probable initiator in a viral marketing campaign. In the later part, a new metric for evaluating social network users is introduced.

4.1 Existing Centrality Measures for Evaluating Users

The centrality measures are commonly used approaches to picking up information initiators for applications that include viral marketing and recommendation systems. In this section, popular centrality approaches are discussed. The most popular centrality measures to measure the importance of a node are degree centrality, closeness centrality, betweenness centrality [43]. The degree centrality assumes that a node that has many direct connections is at the center of the network and plays an important role in information spread. In the context of social networks, degree centrality represents the number of contacts of a user. The closeness centrality focuses on how close a node is to all other nodes in the network. This metric refers to the number of friends separating two individuals. An individual may be linked to a larger portion of the network through a few popular direct friends. This individual himself may have a small degree centrality. The betweenness centrality assumes that if a node is more frequently in the shortest paths between other nodes, it is more important to the network. This metric indicates the power to forward or delay requests between two unfamiliar individuals. Eigenvector centrality is the other metric for measuring a node's popularity in a network. A node's eigenvector centrality is proportional to the sum of the eigenvector centralities of all nodes directly connected to it [12]. This metric indicates the popularity of an individual in the context of social networks.

There are also other metrics such as PageRank [39], hyperlink-induced topic search(HITS) [28], Birnbaum's component importance (BCI) [1] that rank the nodes individually based on their importance. In their basic form, PageRank and HITS value a node merely according to the graph topology [59]. The concept of a hub is prevalent in identifying key users. Users who are in a hub position are characterized by a great potential for communication and interaction within a network[19]. However, in the real-world networks, users who are connected to the most number of users do not show high interaction rates. The concept of the hub also fails to understand the diffusion mechanism.

Centrality measures are suitable for identifying initial information propagators in typical computer networks. In these networks, every receiving node functions as a sender to all its neighbors, those that meet the stated conditions. However, in a social network, it is more of an individual's choice to spread information to certain neighbors. Therefore, these centrality measures may not be suitable neither for evaluating users nor for selecting initiators for information diffusion in social networks.

4.2   Interaction Rate as a Metric to Evaluate Users

In this section, a new metric to evaluate users in a social network is discussed. To understand the need to introduce this new metric, four standard datasets High energy physics (HEP),[1] Physics -Theory (PHY) (see footnote 1), Wikivote (see footnote 1) and YouTube,[2] whose description is in Table 1, are analyzed. For the HEP and PHY datasets, interactions were not available and were synthesized on a power law distribution pattern, which can be produced using MATLAB or a similar tool.[3] The pseudocode for synthesizing such data is in Algorithm 1. The function $randpower(1, n)$ generates n random numbers on power law distribution. To understand the role of users in the network, their friend count, popularly known

---

[1]https://www.microsoft.com/en-us/research/people/weic/#publications

[2]http://socialcomputing.asu.edu

[3]http://tuvalu.santafe.edu/~aaronc/powerlaws

Table 1 Dataset description

Name	No. of nodes	No. of edges	No. of interactions
High energy physics	15,233	58,891	588,136
Physics-theory	37,154	231,584	2,315,840
YouTube	15,088	76,765	2,239,440
Wikivote	8275	103,689	1,057,868

Input: List power, social graph G(V,E)
Output: List powerlist
Initialize $power = \emptyset$, $powerlist = \emptyset$;
$power\{\}$ = randpower(1,n)// where $|V| = n$
for $i = 0 to i < n$ **do**
| $edgelist[i] = e(u_i, v_i)$
end
for $each\ r \in power$ **do**
| $powerlist = powerlist \cup edgelist[r]$
end

Algorithm 1: Pseudocode to synthesize interactions

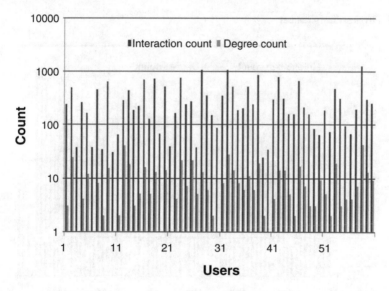

Fig. 2 High energy physics dataset

as "degree of the node", and their interaction count are analyzed. Figures 2, 3, 4, and 5 show the degree count and interaction count of users for the chosen datasets.

When the activities of the users were analyzed, it was observed that a very large portion of the users did not actively take part in the network activities. Instead, a very small portion of this network was involved in these activities. To this end, it is evident that a high number of interactions come from users who maintained a low friend count (small degree). Also, a striking observation can be made that

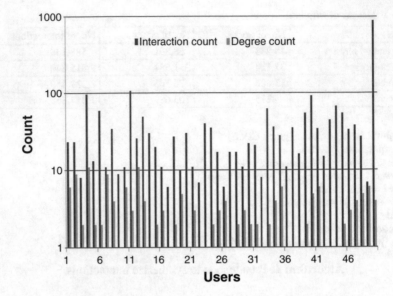

Fig. 3 Physics-theory dataset

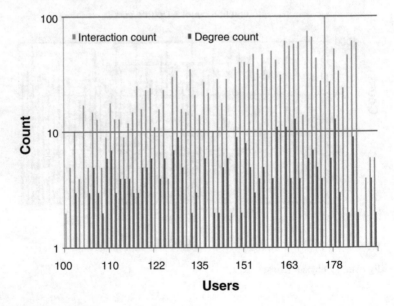

Fig. 4 YouTube dataset

the users with a high degree did not interact well enough among their friends. This observation can also be seen in various other content-based social networks. Conclusively, the traditional approach of evaluating a node with regard to its degree (as discussed in the previous subsection) would not produce accurate results. On

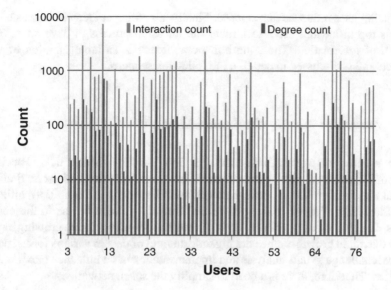

Fig. 5 Wikivote dataset

account of this observation, the interaction count of the users is considered to be an appropriate evaluation metric for applications that rely on user involvement, such as viral marketing.

5 Solving Influence Maximization in a Holistic Approach

Interactions among the users is an important attribute of the network. As most of the social networks, blogs, and forums are content-based models, interactions among users cannot be overlooked. Active users play a vital role in the spread of information or marketing. Based on this premise, a solution to influence maximization is developed that is based on user involvement.

The previous works solve influence maximization either by heuristics or by estimating parameters such as influence. However, the spread of information depends mainly on three factors: network structure, user influence, and seeding strategy. These aspects are explored and a three-stage solution is proposed to solve influence maximization. The growing size of social networks is a major hindrance to analyzing the effectiveness of any algorithm in general. The complexity of an NP-hard problem such as influence maximization increases drastically with an increase in the size of the social network. Therefore, in the first stage of the solution, the scalability issue is tackled by pruning the social network to ascertain the real contributors in the diffusion process. In the second stage, as peer influence is a major factor in the adoption of information or products in social networks, an approach to estimating user influence is proposed. Finally, a seeding strategy is suggested that selects top influential users to initiate the diffusion process to have an effective spread of information. Thus, the holistic approach is an amalgamation of these aspects and is discussed in detail in the following sections.

5.1 Stage 1: Pruning the Social Network

Popular social networking sites such as Google+, Friendster, Flickr, Facebook, Yahoo, Twitter, etc., have grown from a few users to billions of users. Statistics reveal that the number of social network users has increased from 0.97 billion in 2010 to 1.82 billion in 2014 [48]. These numbers are sure to rise in the coming years, clearly showing evidence for the fact that social networks are growing rapidly. With this rapid growth, comes the gigantic amount of data in various forms, posing a great challenge to data analysis and implementation of an influence maximization solution. Therefore, there is a need to simplify the social networks.

5.1.1 Existing Network Simplification Approaches

Various network pruning strategies, such as to maintain connectivity[60], shortest path[58], source to link flow[35], triangular inequality [42], modularity [2], and cut sparsifiers [13] are seen in the literature. Serrano et al. [47] and Foti et al. [11], focus on weighted networks and select edges that represent statistically significant deviations with respect to a null model. An application of a pruning process to connectivity constraint is also seen in [33].

Although a large amount of literature is available on network simplification, these approaches are not suitable for simplifying the social network. There is a possibility of a decline in accuracy, with the increasing erroneous removal of nodes and edges[3]. In most cases, previous works use the structural properties of the graph during the pruning process without understanding whether a link is used for communication. Removing a connection edge from the social graph may lead to disturbance of its structural properties. Therefore, it is important that any metric sought out for pruning the social graph at the edge and node level should retain the properties required for efficient information propagation.

5.1.2 New Approach for Social Network Simplification

The social network is represented as graph $G(V, E)$, where V is the set of users and E is the set of edges that defines the underlying relationship. A link formed between the users is not a random link and indicates that two users are well connected in terms of similar interests and ideas. It is observed that a few of these links are used more often than others and these are the strong links that keep two individuals firmly

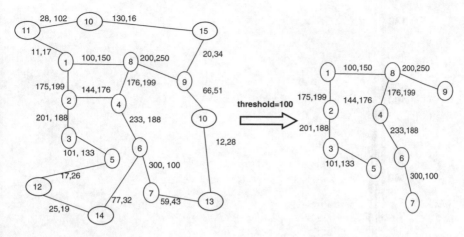

Fig. 6 Pruning of social graph

connected. These individuals have a greater than average potential to influence each other than any two randomly selected users. Therefore, there is a need to distinguish between contact edges and interaction edges. An immediate question to be answered is whether there is a need to keep all those contact edges that have been used at least once for communication. If it is decided to keep all these edges that have been used at least once, the pruning process will not be beneficial. Moreover, there may be a particular pair of users who have not been interacting or who have had very few interactions for a long time. In such a case, retaining this connection is not beneficial. Therefore, there is a need to find the minimum number of interactions needed to classify the user as a contributor or not. A threshold parameter, say *minimum interaction rate*, will remove all those contact edges that have very few interactions in a certain given time window. This threshold prunes the social network to its valuable components. Figure 6 shows the simplification of the original social graph $G(V, E)$ to produce a pruned graph $G_c(V_c, E_c)$.

Interactions play an important role in the dissemination of information. It is observed that every social network has one interaction that stands out over the other interactions. On Facebook, activity is the number of posts, on Twitter it is the number of tweets, on Flicker it is the photo uploads and on YouTube it is the uploads and downloads of videos. Determining the minimum activity rate of users is an open question. For this purpose, the distribution patterns of the interactions of the users in the networks are analyzed. These distribution patterns are shown in Figs. 7, 8, 9, and 10. These distribution patterns follow power law distribution [8]. When any distribution takes such a pattern, the statistical dispersion method of the central tendency, such as mean, mode and standard deviation, cannot be used to get the measure of dispersion. A more robust technique is required to be able to obtain a useful measure of dispersion in the presence of outliers. In the power law distribution pattern, the mean is much larger than the median and the mode, i.e., mode<median<mean. Hence, the mean, cannot be used as the measure of dispersion.

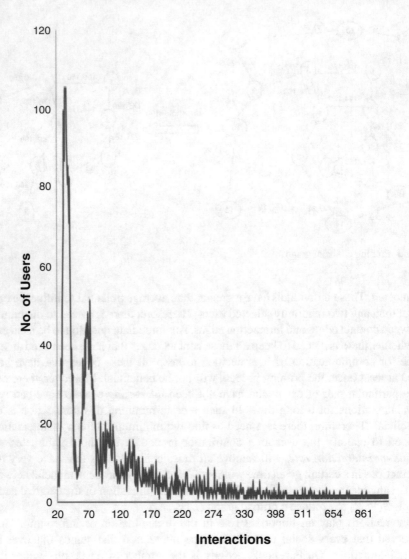

Fig. 7 HEP

For such a distribution, a robust dispersion measure would be median absolute deviation (MAD) [40]. The median absolute deviation approach effectively discards the outliers in the data compared with the standard deviation. Thus, it avoids the need to specifically remove outliers, making the approach less time consuming.

To compute the median absolute deviation, the median for a given population is determined first. Next, the absolute value of the distance between each separate observation and the median is computed. Finally, the median absolute deviation is obtained by computing the median of the values computed in the previous step.

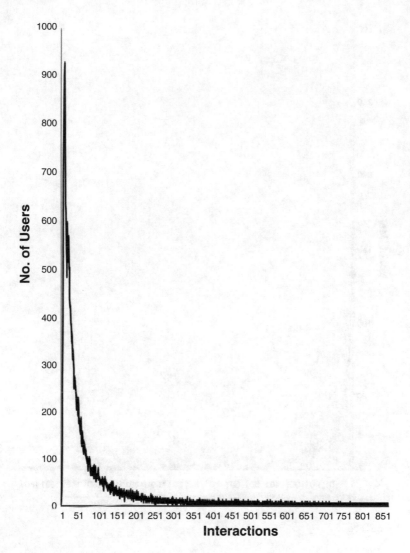

Fig. 8 PHY

More precisely, for a univariate data set $X_1, X_2, \ldots, X_n$, the MAD is defined as the median of the absolute deviations from the median is given as in Eq. (1).

$$\text{MAD} = M(|x_i - M(x_j)|), \tag{1}$$

Social network user data are usually grouped where a value is repeated many times. For example, in the HEP dataset, there are 38,757 users who have interacted only once. Therefore, the frequency of one interaction is 38,757. Such a pattern is seen in almost all the social networks. This is a characteristic feature of power

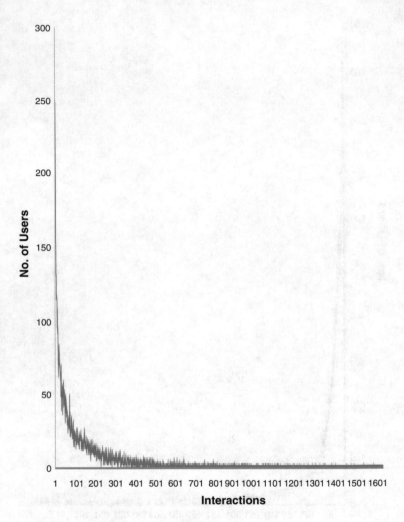

Fig. 9 YouTube

law distribution. For such datasets, the formula for calculating MAD is as given in Eq. (2)

$$MAD = M(f_i * (|x_i - M(x_i)|)), \qquad (2)$$

In the context of the interaction, this would be the ideal count of interactions that would be used to identify a contributor from the rest of the network. As long as the distribution pattern matches the power law distribution, the choice of choosing MAD to define the minimum activity rate is justified.

The approach proposed here reflects the dynamic nature of the social network. Users who may have a high number of interactions within the chosen time period

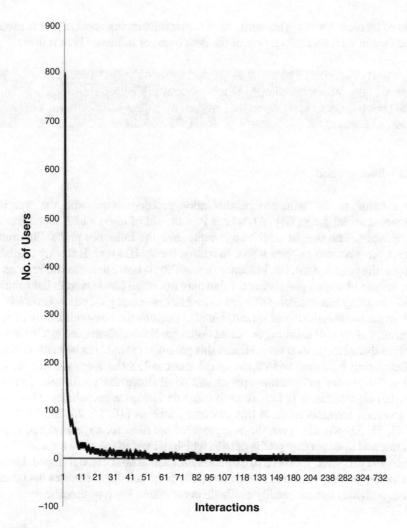

Fig. 10 Wikivote

are included in the pruned graph. By varying the granularity of the time period a close look at the interaction pattern in the social network can be obtained. Details of this approach are available in our previous work [50].

5.2 Stage 2: Estimating User Influence

The importance of influence is quite evident in viral marketing. There are many instances in which a group of people buy products because they were recommended to do so by their friends. This attribute of a user influencing another user is used to advantage in viral marketing. One of the definitions of influence is as follows.

> Social influence is the change in an individual's thoughts, feelings, attitudes, and behaviors that results from interaction from other people or group [31].

5.2.1 Background

The solution to the influence maximization problem starts with the weighted undirected social graph $G(V, E)$, where V is the set of users and E represents the set of edges. The weight on the edge represents the influence probability and is given a pre-assumed uniform value. In reality, the social graph is readily available, whereas the edge weight, i.e., influence probability, is not. There are two reasons for this, the use of such a pre-assumed value may not be an ideal setup in the solution. First, assuming information diffusion to be uniform among all contacts results not only in the overestimation of spread, but also leads to the non-optimal selection of influential users [54] resulting in biased outcome. Second, influence is a behavioral attribute that changes over time. Hence, this parameter should not be made constant.

Estimating influence and obtaining the most influential users are not separate issues. To predict information spread and to evaluate the performance of seed selection algorithms, it is important to estimate influence probability. There have been several attempts made in this direction, such as [10, 16, 20, 23, 27, 29, 33, 44, 45, 51, 53–56]. However, these approaches are resource-expensive and require accurate and in-depth user profile details, which in most of the cases are unavailable. Hence, the proposed approach to estimating influence has been developed, keeping in mind the privacy concerns of the users. The proposed approach uses data related to user activities that are readily available in the action log repositories.

5.2.2 Influx: An Approach to Estimating User Influence

In a social network, various interactions such as posts, likes, recommendations, etc., are seen. In the proposed approach, the type of interaction is not considered, but instead the focus is on knowing whether a pair of connected users interact well enough to influence each other. Further, the approach does not distinguish between a positive influence and a negative influence. However, these could be carried out as separate research.

An approach is developed to estimate the probability of influence of a node using the interaction count of a user. Consider a scenario in which a user A has five contacts B, C, D, E, and F. For illustration, let us assume that the numbers of interactions A has with B, C, D, E, and F are 30, 40, 50, 60, and 30 respectively. As the numbers of interactions of A with his neighbors are different, the probability that A can influence his/her neighbors is also different. Hence, the metric of influence probability of a user should include an aspect of user interaction with his/her neighbors.

The strength of an edge reflects the intensity of the interactions through the tie. The strength of the edge is represented as y_{ij}, which is the number of interactions from v_i to v_j. The number of interactions from user v_i to user v_j is not the same as from user v_j to v_i, i.e., $y_{ij} \neq y_{ji}$. Therefore, the probability of influence is not symmetrical either, i.e., $p_{ij} \neq p_{ji}$.

Thus, the probability of influence is quantified using the interaction count of a user. The probability of influence is calculated as in Eq. (3).

$$p_{u,v} = \frac{y_{u,v}}{S_{s=\{n \in N\}} y_{u,s}} \tag{3}$$

where, N is the set of nodes incident on node u and $y_{u,s}$ is the number of interactions of the node u to the incident node n. The normalization process, sets the value of P within the range $(0,1]$, according to the definition of probability. With this approach, in the given example, $p_{A,B}$ is 0.147. Values on other edges are similarly obtained. This scenario is shown in Fig. 11.

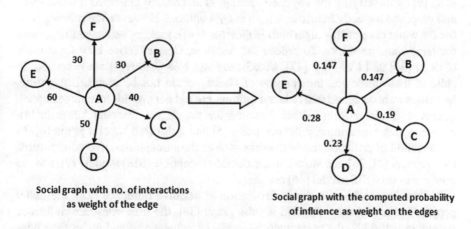

Social graph with no. of interactions
as weight of the edge

Social graph with the computed probability
of influence as weight on the edges

Fig. 11 Social network with influence probability on the edges

5.3 Stage 3: Obtaining the Seed Set

We are now in the position to obtain a good set of initial users, hoping that they will spread the information to a vast population and result in effective sales. In this section, prominent works towards influence maximization are discussed, followed by the proposed approach of ranking users.

5.3.1 Hardness of Influence Maximization

Influence maximization comes under the HP-hard category of problem complexity. Finding a solution in real time, when the input grows exponentially, is impossible for NP-hard problems. Kempe et al. proved that influence maximization under popular diffusion models is NP-hard [24].

Theorem 1 *The influence maximization problem is NP-hard for the independent cascade model and the linear threshold model[24].*

5.3.2 Existing Works

The study of information diffusion in social networks is first proposed by Domingos et al. [9] by identifying the key users. Kempe et al. [24, 25] classified it as NP-hard and proposed a greedy heuristic, which is 66% optimal. However, the running time for the worst case of this algorithm is $O(n^2(m + n))$, making its usage impractical for large scale networks. To reduce the run time, Cost effective lazy forwarding (CELF) [30] and CELF++ [17], MixGreedy and NewGreedy[4] are proposed. In spite of these attempts, the runtime of Greedy could not be reduced; therefore, heuristics were designed that reduced runtime, but did not provide an approximation guarantee. Initial efforts in this direction are the degree discount heuristic [4], coverage under maximum influence paths [5] and a directed acyclic graph [6]. In the direction of diffusion models, works such as the susceptible-infected-recovered (SIR) model [21, 26], the susceptible-infected-susceptible (SIS) model [27], and the continuous time SIS model [46] are seen.

Recent approaches are the incorporation of negative influence [22], the belief propagation model on a directed acyclic graph [38], the time constraint influence spreading paths[32], the combinatorial model of influence spread under time window constraint [15], influence maximization in dynamic networks [61], InFlowMine [49], and the three-step cascade model[41]. Recently, He and Kempe [18] and Chen et al. [7] came up with a new variant for influence maximization, known as robust influence maximization. He and Kempe found the top influential users in the setting in which multiple influence functions are used for the same model [18]. On the other hand, Chen et al. discussed the solution to influence maximization, given an uncertainty in the parameter input. They propose the LUGreedy algorithm to improve the existing greedy algorithm.

5.3.3 User Ranking on Interaction Rate: Outdegree Rank Algorithm

The formalization of influence maximization and the runtime issue of the greedy approach paved the way for various other alternatives. However, several improvements to the original greedy approach are still not efficient. Therefore, heuristics are designed to solve runtime issues, but compromise optimality. Of these, the degree centrality heuristic has proved to be efficient and close to an optimal solution [5]. The degree concept obtains users with the highest degree (a.k.a. number of contacts), with the belief that such a user will trigger a vast outbreak of information, leading to adoption. However, in the real world, a user interacts with only a small percentage of his/her contacts, raising suspicion regarding the viability of the degree heuristic. Also, other variations of the degree heuristic arouse similar concerns. To make the degree concept viable in the real world, the outdegree rank heuristic for influence maximization is proposed in this work. Unlike existing works, the outdegree rank heuristic considers the user attribute, i.e., interaction count, to obtain the most influential users.

Before getting into the details of the outdegree rank heuristic, three terms: contact degree, interaction graph and interaction degree are introduced to understand the concept.

Definition 1 For a node v, its contact degree is referred to in the social network graph as the number of edges incident on it and is denoted as $Cd(v)$.

Definition 2 The interaction graph is a multi-edge social graph of the contributor graph, where edges represent the interaction between each pair of nodes.

Definition 3 For a node v, its interaction degree is referred to in the interaction graph as the number of edges incident on it and is denoted as $Id(v)$.

User activity types are unique to every social network, for example, wall posts, comments, likes, following, status updates, etc. However, not all users may be actively involved in these activities. Most of the content usually comes from a very small percentage of network users. In a scenario where a user v maintains interactions among few friends, $Id(v) \ll Cd(v)$. Therefore, in this case, the degree centrality approach, which uses $Cd(v)$ as a metric to rank users, may not be optimal.

In this view, degree centrality needs further investigation and the solution is explored in the interaction graph.

It is true that there does not exist any proven correlation between the degree of the node, $Cd(v)$, and interaction degree, $Id(v)$, but an attempt to formalize a relation is made here based on the following reason. When a user's interaction degree is very large compared with his degree, it can be concluded that he/she is interacting actively with at least a few of his/her friends. Also, if $Id(v)$ is very much less than $Cd(v)$, it can safely be concluded that the user is not interacting with all his/her contacts. With $Id(v)$ broken into outdegree and indegree, the ranking of users can be further improved.

The $Id(v)$ can be further specified in terms of *indegree*, defined as the number of edges leading to that node and *outdegree*, defined as the number of edges leading away from that node. In an interaction graph, the $Indegree(v)$ represents the popularity index and $Outdegree(v)$ represents the participation index. The concept of indegree is already explored in the PageRank heuristic for finding popular web pages [39] and identifying key users in social networks [19]. To solve influence maximization, the final stage is aimed to ranking users on their outdegree.

5.3.4 Outdegree Rank Heuristic for Influence Maximization

For a node, if $outdegree(v) \gg Cd(v)$, it shows that node v is highly interactive. Such a node has greater potential to spread information in the network than other nodes. Based on this surmise, the outdegree(v) is used to rank the users in the social network. This approach is referred to as the outdegree rank heuristic. Unlike the degree heuristic, in which the $Cd(v)$ is almost static, the outdegree(v) is frequently varying according to the changes in the interaction rate of the user. In this way, the outdegree rank reflects the dynamic changes occurring in the social network. Thus, the outdegree rank is a viable solution in the real world.

5.4   Algorithm

Algorithm 2 is a consolidated algorithm for the three steps discussed previously. The input to the algorithm is the social network $G(V, E)$, activity log A, activity record AR, which maintains the number of activities of each user, and k, the number of seeds.

Input: G(V,E), Activity log $A(u, v)$, $AR(u, count)$,k
Output: Seed set S of k nodes
Initialize $u.count = 0$, $G_c(E_c, V_c) = \emptyset$, $Total_{u,s} = 0$, $S = \emptyset$;
for *Each u in activity log A(u, v)* **do**
   | **if** *u.isadded = FALSE* **then**
   | | u.isadded=TRUE;
   | **end**
   | u.count++;
end
Update $AR(u, count)$
Compute the threshold *alpha* using Eq. (2)
for *Each edge e(u, v) $\in$ E* **do**
   | **if** *u.count $\geq$ alpha && v.count $\geq$ alpha* **then**
   | | $E_c = E_c \bigcup e$
   | **end**
end
for *each e(u, v) $\in$ E_c* **do**
   | S=neighbourset(u)
   | **for** *each s $\in$ S* **do**
   | | $Total_{u,s} + = y_{u,s}$
   | **end**
   | $p(u, v) = \frac{y_{u,v}}{Total_{u,s}}$
end
for *each vertex v $\in$ V_c* **do**
   | compute its outdegree out_v
end
for *i=1 to k* **do**
   | $u - argmax_v\{out_v | v \in V_c - S\}$
   | S= S $\cup$ u
end
Output S

Algorithm 2: Influence maximization algorithm

6 Results

6.1 Structure Pruning to Find Contributors

The original social networks those described in Table 1, are pruned to fetch the contributor graph whose description is shown in Table 2. The results show that the pruned graph is now scalable to analyze the effectiveness of influence maximization algorithms. Further, the pruned network should have information propagation properties. To analyze these properties, the pruned network is evaluated on small world metrics, which includes, a higher clustering coefficient, a smaller diameter, smaller average path lengths, lower modularity and fewer components. Tables 3 and 4 summarize the results. The pruned network is compared with the original social network and with the pruned graph obtained by removing those edges that do not constitute the shortest paths. The pruned graph obtained by the proposed method exhibits small world properties significantly better. Thus, the pruned graph

Table 2 Value of *alpha*, V_c and E_c of $G_c(V_c, E_c)$

Name	$G(V)$	$G(E)$	alpha	$G_c(V_c)$	$G_c(E_c)$
HEP	15,233	58,891	403	205	405
PHY	37,154	231,584	637	310	2934
YouTube	15,088	76,765	1544	545	4846
Wikivote	8275	103,689	660	410	11,726

Table 3 Structural properties of the social graph and contributor graph of HEP and PHY

Attribute	HEP			PHY		
	Original graph	Shortest path method	Proposed approach	Original graph	Shortest path method	Proposed approach
Nodes	15,233	7159	205	37,154	37,149	310
Edges	58,891	25,914	405	23,1584	174,161	2934
ACC	0.261	0.253	0.189	0.403	0.7	0.482
Diameter	31	19	7	23	19	11
APL	10.137	5.424	2.33	7.63	6.259	3.736
Components	1779	4662	9	3878	3456	14
Modularity	0.852	0.712	0.722	0.927	0.923	0.7

Table 4 Structural properties of the social graph and contributor graph of YouTube and Wikivote

Attribute	YouTube			Wikivote		
	Original graph	Shortest path method	Proposed approach	Original graph	Shortest path method	Proposed approach
Nodes	15,233	7159	205	37,154	37,149	310
Edges	58,891	25,914	405	23,1584	174,161	2934
ACC	0.261	0.253	0.189	0.403	0.7	0.482
Diameter	31	19	7	23	19	11
APL	10.137	5.424	2.33	7.63	6.259	3.736
Components	1779	4662	9	3878	3456	14
Modularity	0.852	0.712	0.722	0.927	0.923	0.7

is an ideal substitute for the original social graph with regard to applications that are expensive in terms of resource usage. The graph size in terms of edges and nodes is drastically reduced, yet the properties the of pruned graph are well suited to information propagation compared with those of the original social network and can thus can be used in place of the original network in applications such as viral marketing.

6.2 Estimating User Influence

In this section, the impact of using an estimated influence probability on the spread is discussed. Influence probabilities developed from various other approaches are used to predict spread in various setups. The results are then compared to demonstrate the effectiveness of the proposed approach. The following four cases are used to predict information spread.

1. Trivalency model (TVM) where $p = 0.01$.
2. Random numbers with uniform distribution probability(RNUDp): The influence probabilities are generated from a uniform distribution.
3. Random numbers with normal distribution probability (RNNDp): The influence probabilities are generated from a normal distribution.
4. Influx.

The diffusion process is observed under an independent cascade model run for 1000 simulations for accuracy. The model uses the probability of influence obtained from the above approaches and the seed set size of 50 is considered, which are obtained from the standard algorithms for the seed selection process for influence maximization found in the literature such as highest degree [9], distance [9], single discount [4], and degree discount [4], and are used to compare the outcomes of the proposed approach. The spread estimate is measured as the count of the number of nodes that are activated at the end of the process. The results are shown in Figs. 12, 13, 14, and 15 respectively.

For the HEP dataset, the results, when compared with the approaches that use RNNDp and RNUDp, show an increase of 20% and 18% for the highest degree, 15% and 14% for the distance heuristic, 27.4% and 24.75% for the single discount, and 30% and 26.8% for the degree discount. Moreover, when the value of influence is fixed at 0.01, the influx approach shows an increase of 20%, 15.6%, 28.5%, and 30% for the highest degree, distance, single discount, and degree discount approaches respectively.

For the PHY dataset, too, there is an increase of 8.9%, 14%, 18.43% and 23.17% compared with RNNDp for the highest degree, distance, single discount, and degree discount approaches respectively, and 4.8%, 8.9%, 9.8%, and 12.6% compared with RNUDp for each of the highest degree, distance, single discount, and degree discount approaches respectively. In addition, there is an increase of 8.6%, 14%, 20%, and 25.53% compared with the TVM approach for each of the highest degree, distance, single discount, and degree discount approaches respectively.

For the YouTube dataset, there is also an increase of 9.6%, 15.7%, 10.5% and 10.6% compared with RNNDp for the highest degree, distance, single discount, and degree discount approaches respectively, and 3.15%, 6.3%, 6.3%, and 6.1% compared with RNUDp for the highest degree, distance, single discount, and degree discount approaches respectively. In addition, there is an increase of 19%, 24.21%, 22.1%, and 19.6% compared with the TVM approach for each of these approaches respectively.

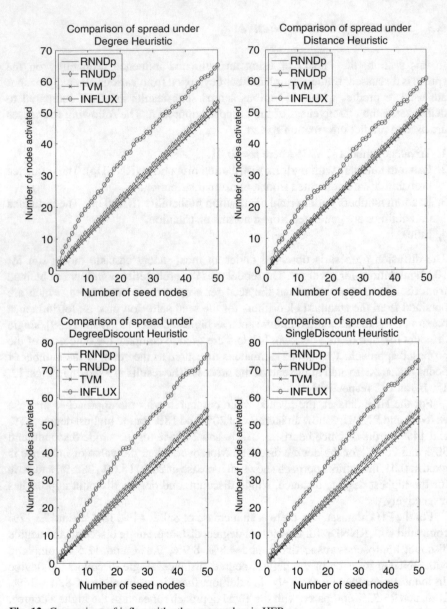

Fig. 12 Comparison of influx with other approaches in HEP

Finally, for the Wikivote dataset, there is an increase of 15%, 13.15%, 13.2% and 15.11% compared with RNNDp for the highest degree, distance, single discount, and degree discount approaches respectively and 40.4%, 35.52%, 39.7%, and 40.6% compared with RNUDp for the highest degree, distance, single discount, and degree discount approaches respectively. In addition, there is an increase of 34.47%, 28%,

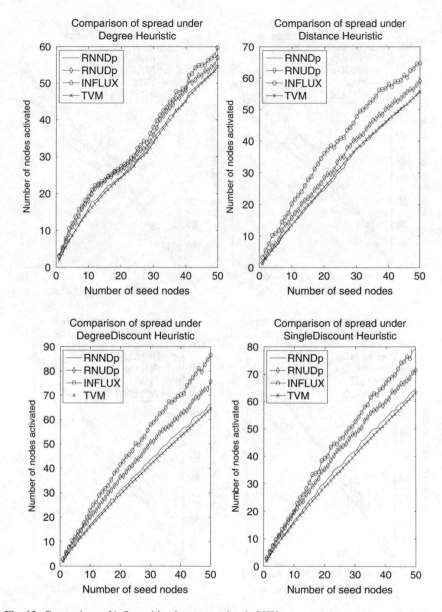

Fig. 13 Comparison of influx with other approaches in PHY

33.26%, and 35.8% compared with the TVM approach for each of these approaches respectively.

Overall, the information spread obtained by using the influx approach is 5–30% higher than RNNDp, RNUDp, and pre-assumed $p = 0.01$ for each of the standard algorithms such as highest degree, distance, single discount, and degree discount. It

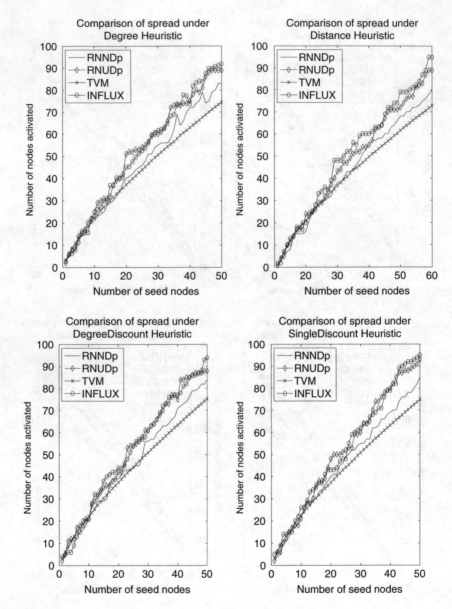

Fig. 14 Comparison of influx with other approaches in YouTube

is clear that the influx approach yields a better outcome as it reflects the interactive nature of the users. whereas the existing approach of using the TVM does not account for user inclination when predicting the spread.

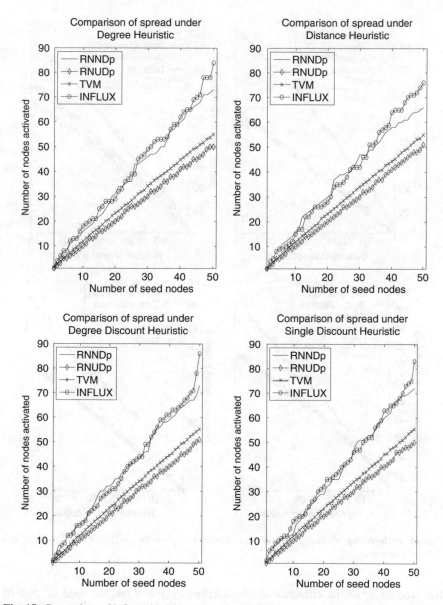

Fig. 15 Comparison of influx with other approaches in Wikivote

6.3 Outdegree Rank with Estimated Influence

Finally, in this section, the performance gain obtained using the outdegree rank with influence estimate (ORIE) is highlighted. The diffusion process is observed under an independent cascade model run for 1000 simulations for accuracy. The model uses

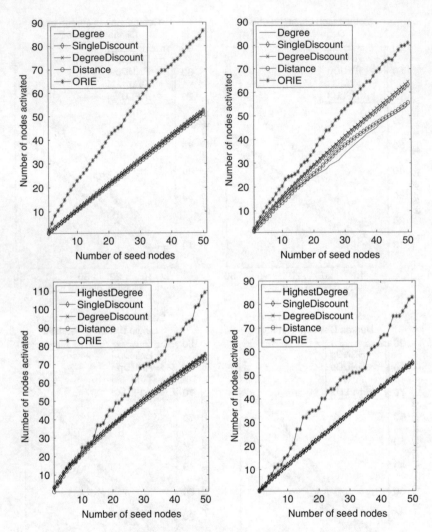

Fig. 16 Performance of outdegree rank with influence estimate on the HEP, PHY, YouTube, and Wikivote datasets

the probability of the influence-obtained influx approach and the seed set size of 50 is considered, which is obtained from standard algorithms used in Sect. 5.2. The spread estimate is measured as the count of the number of nodes that are activated at the end of the process. The performance gain of ORIE compared with other state-of-the-art (SOA) approaches is shown in Fig. 16.

In the HEP dataset, there is an increase of 40.5% compared with the degree, 39.47% compared with the degree single discount, 39.2% compared with the degree discount, and 41% compared with the distance heuristics. Similarly, in the PHY dataset, there is an increase of 32.6% compared with the degree, 21.9% compared with the degree single discount, 20.5% compared with the degree discount, and 31.5% compared with the distance heuristics. In the YouTube dataset, there is an increase of 31.7% compared with the degree, 31.2% compared with the degree single discount, 30.8% compared with the degree discount, and 33.1% compared with the distance heuristics. Finally, in the Wikivote dataset, there is an increase of 33.6% compared with the degree, 33.2% compared with the degree single discount, 33.4% compared with the degree discount, and 33.8% compared with the distance heuristics.

With the new ranking strategy, there is a gain of up to 41% in information spread compared with the SOA approaches. Conclusively, the outdegree rank strategy outperforms standard approaches that are based on degree centrality.

7 Conclusion

The aim of the chapter is to introduce a new perspective with regard to solving influence maximization. The influence maximization and information diffusion processes are like two faces of the same coin, and should be studied in each other's context. In this chapter, a holistic, an effective, and a scalable solution to influence maximization is proposed. The influence maximization solution is achieved under a combination of aspects of structure, heuristic, and user influence. The work presents three novel approaches to these three aspects, namely structure pruning, influx to estimate influence, and outdegree rank. Finally, the amalgamation of these aspects is a contribution to influence maximization. The experiments on various cases support the claim that user attributes determine the diffusion process; thus, the approach contributes to a new direction of influence maximization solutions. As a future work, the three aspects can be dealt with in detail, adding more features. When the constraints on the privacy of user data for applications is flexible, more user features can be used in each of the phases to provide a better working model. Also, another direction is verifying the validity of the proposed approaches on other diffusion models. Although this work is based on the surmise that all users are equally popular, it can however be extended to include the celebrity aspect.

References

1. Andrews, J.D., Beeson, S.: Birnbaum's measure of component importance for noncoherent systems. IEEE Trans. Reliab. **52**(2), 213–219 (2003)
2. Arenas, A., Duch, J., Fernandez, A., Gomez, S.: Size reduction of complex networks preserving modularity. CoRR (2007). abs/physics/0702015
3. Borgatti, S.P., Carley, K.M., Krackhardt, D.: On the robustness of centrality measures under conditions of imperfect data. Soc. Netw. **28**(2), 124–136 (2006)
4. Chen, W., Wang, Y., Yang, S.: Efficient influence maximization in social networks. In: Proceedings of the 15th ACM SIGKDD, KDD '09, pp. 199–208. ACM, New York, NY (2009). doi:10.1145/1557019.1557047

5. Chen, W., Wang, C., Wang, Y.: Scalable influence maximization for prevalent viral marketing in large-scale social networks. In: Proceedings of the 16th ACM SIGKDD, KDD '10, pp. 1029–1038. ACM, New York, NY (2010) doi:10.1145/1835804.1835934

6. Chen, W., Yuan, Y., Zhang, L.: Scalable influence maximization in social networks under the linear threshold model. In: Proceedings of the 2010 IEEE International Conference on Data Mining, ICDM '10, pp. 88–97 (2010)

7. Chen, W., Lin, T., Tan, Z., Zhao, M., Zhou, X.: Robust influence maximization. CoRR (2016). abs/1601.06551. http://arxiv.org/abs/1601.06551

8. Clauset, A., Shalizi, C.R., Newman, M.E.J.: Power-law distributions in empirical data. SIAM Rev. **51**(4), 661–703 (2009). doi:10.1137/070710111. http://dx.doi.org/10.1137/070710111

9. Domingos, P., Richardson, M.: Mining the network value of customers. In: Proceedings of the Seventh ACM SIGKDD, KDD '01, pp. 57–66. ACM, New York, NY (2001). doi:10.1145/502512.502525

10. Fang, X., Hu, P.J.H., Li, Z., Tsai, W.: Predicting adoption probabilities in social networks. Inf. Syst. Res. **24**(1), 128–145 (2013)

11. Foti, N.J., Hughes, J.M., Rockmore, D.N.: Nonparametric sparsification of complex multiscale networks. PLoS One **6**(2), 16431 (2011). doi:10.1371/journal.pone.0016431

12. Freeman, L.C.: Centrality in social networks conceptual clarification. Soc. Netw. **1**(3), 215–239 (1978)

13. Fung, W.S., Hariharan, R., Harvey, N.J.A., Panigrahi, D.: A general framework for graph sparsification. In: Fortnow, L., Vadhan, S.P. (eds.) STOC. pp. 71–80. ACM, New York, NY (2011)

14. Ganesan, K.: Case study on ripple effects of ice bucket challenge on social media channels (2016). http://www.digitalvidya.com/blog/

15. Gargano, L., Hell, P., Peters, J., Vaccaro, U.: Influence diffusion in social networks under time window constraints. In: Structural Information and Communication Complexity: 20th International Colloquium, SIROCCO 2013, Ischia, July 1–3, 2013. Revised Selected Papers

16. Goyal, A., Bonchi, F., Lakshmanan, L.V.: Learning influence probabilities in social networks. In: Proceedings of the Third ACM International Conference on Web Search and Data Mining, WSDM '10, pp. 241–250. ACM, New York, NY (2010)

17. Goyal, A., Lu, W., Lakshmanan, L.V.: CELF++: optimizing the greedy algorithm for influence maximization in social networks. In: Proceedings of the 20th International Conference Companion on World Wide Web, WWW '11, pp. 47–48. ACM, New York, NY (2011). doi:10.1145/1963192.1963217

18. He, X., Kempe, D.: Robust influence maximization. CoRR (2016). abs/1602.05240 http://arxiv.org/abs/1602.05240

19. Heidemann, J., Klier, M., Probst, F.: Identifying key users in online social networks: a pagerank based approach. In: Sabherwal, R., Sumner, M. (eds.) ICIS, p. 79. Association for Information Systems (2010)

20. Jiang, J., Wilson, C., Wang, X., Sha, W., Huang, P., Dai, Y., Zhao, B.Y.: Understanding latent interactions in online social networks. ACM Trans. Web **7**(4), 18 (2013)

21. Johnson, T.: Mathematical modeling of diseases: susceptible-infected-recovered (sir) model (2009). http://op12no2.me/stuff/tjsir.pdf

22. Jung, K., Heo, W., Chen, W.: IRIE: a scalable influence maximization algorithm for independent cascade model and its extensions. CoRR (2011). abs/1111.4795

23. Kasthurirathna, D., Harre, M., Piraveenan, M.: Influence modelling using bounded rationality in social networks. In: Proceedings of the 2015 IEEE/ACM International Conference on Advances in Social Networks Analysis and Mining 2015, pp. 33–40. ACM, New York, NY (2015)

24. Kempe, D., Kleinberg, J., Tardos, E.: Maximizing the spread of influence through a social network. In: Proceedings of the Ninth ACM SIGKDD, KDD '03, pp. 137–146. ACM, New York, NY (2003). doi:10.1145/956750.956769

25. Kempe, D., Kleinberg, J., Tardos, E.: Influential nodes in a diffusion model for social networks. In: Proceedings of the 32Nd International Conference on Automata, Languages and Programming, ICALP'05, pp. 1127–1138. Springer, Berlin, Heidelberg (2005)

26. Kermack, W.O., McKendrick, A.G.: Contributions to the mathematical theory of epidemics. ii. The problem of endemicity. Proc. R. Soc. Lond. **138**(834), (1932). doi:10.1098/rspa.1932.0171

27. Kimura, M., Saito, K., Motoda, H.: Efficient estimation of influence functions for sis model on social networks. In: Proceedings of the 21st International Joint Conference on Artificial Intelligence, IJCAI'09, pp. 2046–2051. Morgan Kaufmann, San Francisco, CA (2009)

28. Kleinberg, J.M.: Authoritative sources in a hyperlinked environment. J. ACM **46**(5), 604–632 (1999)

29. Kutzkov, K., Bifet, A., Bonchi, F., Gionis, A.: Strip: stream learning of influence probabilities. In: Proceedings of the 19th ACM SIGKDD, KDD '13, pp. 275–283. ACM, New York, NY (2013)

30. Leskovec, J., Krause, A., Guestrin, C., Faloutsos, C., VanBriesen, J., Glance, N.: Cost-effective outbreak detection in networks. In: Proceedings of the 13th ACM SIGKDD, KDD '07, pp. 420–429. ACM, New York, NY (2007)

31. Lisa, R.: Social influence. In: The Blackwell Encyclopedia of Sociology, pp. 4426–4429. Oxford Blackwell, Malden, MA (2008)

32. Liu, B., Cong, G., Xu, D., Zeng, Y.: Time constrained influence maximization in social networks. In: 2012 IEEE 12th International Conference on Data Mining (ICDM), IEEE, pp. 439–448 (2012)

33. Mathioudakis, M., Bonchi, F., Castillo, C., Gionis, A., Ukkonen, A.: Sparsification of influence networks. In: Proceedings of the 17th ACM SIGKDD, KDD '11, pp. 529–537. ACM, New York, NY (2011)

34. McCracken, G.: How ford got social marketing right (2010). https://hbr.org/2010/01/ford-recently-wrapped-the-firs/

35. Misiolek, E., Chen, D.Z.: Two flow network simplification algorithms. Inf. Process. Lett. **97**(5), 197–202 (2006)

36. Mullaney, T.: Social media is reinventing how business is done (2012). http://www.usatoday.com/money/economy/story/2012-05-14/social-media economy-companies/55029088/1/

37. Myerson, R.: Graphs and cooperation in games. In: Dutta, B., Jackson, M. (eds.) Networks and Groups, Studies in Economic Design, pp. 17–22. Springer, Berlin, Heidelberg (2003). doi:10.1007/978-3-540-24790-6_2

38. Nguyen, H., Zheng, R.: Influence spread in large-scale social networks–a belief propagation approach. In: Machine Learning and Knowledge Discovery in Databases, pp. 515–530. Springer, Berlin (2012)

39. Page, L., Brin, S., Motwani, R., Winograd, T.: The pagerank citation ranking: bringing order to the web. Standford Infolab (1999)

40. Pham-Gia, T., Hung, T.: The mean and median absolute deviations. Math. Comput. Model. **34**(7–8), 921–936 (2001)

41. Qin, Y., Ma, J., Gao, S.: Efficient influence maximization under TSCM: a suitable diffusion model in online social networks. Soft Comput. 1–12 (2016). doi:10.1007/s00500-016-2068-3

42. Quirin, A., Cordn, O., Santamara, J., Vargas-Quesada, B., Moya-Anegn, F.: A new variant of the pathfinder algorithm to generate large visual science maps in cubic time. Inf. Process. Manage. **44**(4), 1611–1623 (2008)

43. **1**, 215–239 (2008) Robert, H.: Applicability of graph metrics when analyzing online social networks. Curr. Issues IT-Manage. **1**, 215–239 (2008)

44. Romero, D.M., Galuba, W., Asur, S., Huberman, B.A.: Influence and passivity in social media. In: Machine learning and knowledge discovery in databases, pp. 18–33. Springer, Berlin (2011)

45. Saito, K., Nakano, R., Kimura, M.: Prediction of information diffusion probabilities for independent cascade model. In: Lovrek, I., Howlett, R., Jain, L. (eds.) Knowledge-Based Intelligent Information and Engineering Systems. Lecture Notes in Computer Science, vol. 5179, pp. 67–75. Springer, Berlin, Heidelberg (2008). doi:10.1007/978-3-540-85567-5_9

46. Saito, K., Kimura, M., Ohara, K., Motoda, H.: Efficient estimation of cumulative influence for multiple activation information diffusion model with continuous time delay. In: PRICAI 2010: Trends in Artificial Intelligence, Daegu, pp. 244–255 (2010)

47. Serrano, M.A., Bog, M., Vespignani, A.: Extracting the multiscale backbone of complex weighted networks. Proc. Natl. Acad. Sci. **106**(16), 6483–6488 (2009)
48. Smith, C.: How many people use the top social media, apps & services (2014). Http://expandedramblings.com
49. Subbian, K., Aggarwal, C., Srivastava, J.: Mining influencers using information flows in social streams. ACM Trans. Knowl. Discov. Data **10**(3), 26:1–26:28 (2016). doi:10.1145/2815625
50. Sumith, N., Annappa, B., Bhattacharya, S.: Social network pruning for building optimal social network: a user perspective. Knowl. Based Syst. **117**, 101–110 (2017)
51. Teng, Y.W., Tai, C.H., Yu, P.S., Chen, M.S.: Modeling and utilizing dynamic influence strength for personalized promotion. In: Proceedings of the 2015 IEEE/ACM International Conference on Advances in Social Networks Analysis and Mining 2015, pp. 57–64. ACM, New York, NY (2015)
52. Treagus, P.: The dark knight: a case study of viral marketing (2014). http://philtreagus.com/the-dark-knight-a-case-study-of-viral-marketing/
53. Wang, Z., Qian, Z., Lu, S.: A probability based algorithm for influence maximization in social networks. In: Proceedings of the 5th Asia-Pacific Symposium on Internetware, Internetware '13, pp. 12:1–12:7. ACM, New York, NY (2013)
54. Wilson, C., Boe, B., Sala, A., Puttaswamy, K.P., Zhao, B.Y.: User interactions in social networks and their implications. In: Proceedings of the 4th ACM European Conference on Computer systems, pp. 205–218. ACM, New York, NY (2009)
55. Xiang, R., Neville, J., Rogati, M.: Modeling relationship strength in online social networks. In: Proceedings of the 19th International Conference on World Wide Web, pp. 981–990. ACM, New York, NY (2010)
56. Yang, J., Leskovec, J.: Modeling information diffusion in implicit networks. In: 2010 IEEE 10th International Conference on Data Mining (ICDM), pp. 599–608. IEEE, New York, NY (2010)
57. Zhang, H., Mishra, S., Thai, M.T., Wu, J., Wang, Y.: Recent advances in information diffusion and influence maximization in complex social networks. Oppor. Mobile Soc. Netw. **37** (1.1) (2014)
58. Zhou, F., Malher, S., Toivonen, H.: Network simplification with minimal loss of connectivity. In: 2010 IEEE 10th International Conference on Data Mining (ICDM), pp. 659–668 (2010). doi:10.1109/ICDM.2010.133
59. Zhou, F., Mahler, S., Toivonen, H.: Review of bisonet abstraction techniques. In: Bisociative Knowledge Discovery, pp. 166–178. Springer, Berlin (2012)
60. Zhou, F., Mahler, S., Toivonen, H.: Simplification of networks by edge pruning. In: Berthold, M.R. (ed.) Bisociative Knowledge Discovery. Lecture Notes in Computer Science, vol. 7250, pp. 179–198. Springer, Berlin (2012)
61. Zhuang, H., Sun, Y., Tang, J., Zhang, J., Sun, X.: Influence maximization in dynamic social networks. In: 2013 IEEE 13th International Conference on Data Mining (ICDM), pp. 1313–1318. IEEE, New York (2013)

Opinion Dynamics Through Natural Phenomenon of Grain Growth and Population Migration

Puja Munjal, Sandeep Kumar, Lalit Kumar, and Aashish Banati

Abstract Opinion dynamics has witnessed a colossal interest in research and development activity aimed at the realization of intelligent systems facilitating the understanding and prediction of these. Many nature-inspired phenomena have been used for modelling and investigation of opinion formation. One of the prominent models based on the concept of ferromagnetism is the Ising model in statistical mechanics. The model represents magnetic dipole moments of atomic spins, which can exist in any one of two states, +1 or −1. We have used NetLogo to simulate the Ising model and correlated the results with opinion dynamics within its purview. For the first time, the grain growth phenomenon has been investigated to analyze opinion dynamics. We have also modelled natural phenomena of population growth using rigorous mathematics and corroborated the results with opinion dynamics. The results substantiate the potential use of such nature-inspired phenomena that encompass an ensemble large enough to be investigated and correlated with the real world, in terms of the involvement of numerous actors/agents participating in the process of opinion formation.

Keywords Grain growth • Ising model • Nature-inspired algorithms • Net Logo • Opinion dynamics • Population model

P. Munjal (✉)
Jagannath International Management School, Vasant Kunj, Delhi 110070, India
e-mail: puja.munjal@gmail.com

S. Kumar
Jagannath University, Jaipur 303901, India
e-mail: sandpoonia@gmail.com

L. Kumar
Department of Physics, Hindu College, University of Delhi, Delhi 110007, India
e-mail: lalitkrchauhan@gmail.com

A. Banati
Scientist 'F', STQC DTE, Ministry of Electronics and Information Technology, Delhi 110003, India
e-mail: aashish.banati@gmail.com

© Springer International Publishing AG 2017
H. Banati et al. (eds.), *Hybrid Intelligence for Social Networks*,
DOI 10.1007/978-3-319-65139-2_7

161

1 Introduction

In recent times, opinion dynamics based social network analysis have intrigued not only social scientists [3, 7, 22], but also physicists [4, 11, 16, 19, 27, 29, 32, 35, 39, 40, 43, 44], mathematicians [14, 20, 23, 28, 34] and computer scientists [1, 2, 5, 9, 12, 13, 15, 18, 24], resulting in large-scale activity in the emerging inter-disciplinary field of complex system science. Most real-world networks, including the World Wide Web, the Internet, basic cellular networks and many others, are complex. Opinion formation is a complex process affected by the interaction of various components, including individual inclination, the influence of positive and negative peer interaction, the information each individual is exposed to and many others.

The vast area of opinion dynamics deals with the formation of various possible responses to the same issue within a given period of time. Emergence of opinions is a huge category within the social spreading phenomenon, which is facilitated by the diffusion of information. At times, information diffusion is a word-of-mouth-like phenomenon and recently this has been facilitated by online social networks. Although considering opinion dynamics using online social networks, the core issue is the factors influencing the rate of opinion formation. The online social networks consist of several complex, mutually influencing and interconnected entities. For such complex systems, computer simulation becomes a beneficial tool for analysis. Several models inspired from those in use in biology, physics, mathematics etc., have been developed understanding and identifying the mechanisms involved in the opinion formation process. The models range from binary simple models, such as the Ising model, the voter model, to multi-dimensional continuous approaches. In the real world, the formation of an individual's opinion is largely affected by his/her nearest neighbours (social group). As the social group with one type of opinion becomes prominent in the vicinity, it largely influences the opinion of an individual. A group of neighbouring agents with the same opinion try to influence and convince all their neighbours to adapt to their opinion. Such collective phenomena are common in nature from macroscopic (flocks of birds) to macroscopic levels (grain growth phenomenon) [6, 8, 25, 38, 41]. External factors such as information in the form of mass media, can play a key role in driving opinion formation at the population level and in online social communities. Thus, study of the effect of multiple external information sources could greatly benefit studies on opinion dynamics.

Computer simulations a play crucial role in the study of social dynamics and one of the most effective practices employed in social dynamics is agent-based mod-elling. The basic idea is to simulate agents analogous to the real-world phenomena. NetLogo is an integrated multi-agent modelling environment. It includes a graphical user interface for exploring, experimenting with and visualizing models, in addition to a multi-agent modelling language (MAML) used for authoring models. Such languages enable users easily to create and operate numerous graphical agents and define simple rules that govern the agents' behaviour. The NetLogo agents can perform simple rule-based behaviours, such as seeking or avoiding being surrounded by other agents. Such simple agent rules, however, give rise to complex emergent aggregate phenomena, many of which are congruent with their traditional macroscopic formula-based descriptions.

The outline of this chapter is as follows. In Sect. 2 we provide a review of the existing models of opinion dynamics. In Sect. 3 the Ising model is simulated through NetLogo. Section 4 correlates grain growth phenomenon with real world opinion formation and have done computer simulation using NetLogo. In Sect. 5 we have developed a model for population dynamics based on the Ising model and simulated the opinion formation. The chapter concludes by highlighting the simulation results and the contribution of the work.

2 Existing Models of Opinion Dynamics

2.1 Ising Model

One of the oldest, most prominent models and extremely simplified agent-based model applied to opinion dynamics from physics is the Ising model [10, 17, 30, 33, 36]. The main attribute of the models based on agent is that individuals are assumed to be independent whose communication with each other results in update of their opinions according to constraints imposed by fixed rules. The agents' interactions may occur either in groups or pair wise, usually between nearest neighbours and are connected by a primary graph defining the topology of the system. In the Ising model, a spin represents the opinion of each agent; based on two opinions, it can be upward or downward. In this model, the peer interaction is represented by spin coupling and the magnetic field corresponds to the external information. In spite of being a very simple model, it is of huge relevance because of its anticipation of a phase transition from an ordered to a disordered phase, associated with the strength of the spin's interaction.

2.2 Voter Model

This model was originally considered for analyzing the competition of species [9, 26]. In this model, the agents do not explicitly imitate their neighbours, but in an average sense. The relevance of this model lies in the fact that it can be solved precisely in any dimension.

2.3 Majority Rule Model

In this model, all agents are supposed to communicate with each other, to represent the social network as a complete graph. It is based on a simple concept of selecting a group of agents in which all agents take the majority opinion inside the group [21].

2.4 Sznajd Model

The concept of the Sznajd model is that the opinion of an individual is influenced more by a group rather than another individual [37]. This model basically describes how the opinion spreads in society.

3 Simulation of the Ising Model Using NetLogo

The models described above are an extension of the Ising model, in one form or another. Therefore, we first performed simulations for the Ising model. We have simulated the Ising model using NetLogo[42] and tried to establish that this model is relevant even today for studying opinion dynamics. The NetLogo model used for the two-dimensional Ising model uses the Monte Carlo simulation based on the Metropolis algorithm. The Metropolis algorithm originates from a paper dating back to 1953, by Nicholas Metropolis et al. [31].

Let two numbers +1 and −1 denote the state of spins. The +1 spins are represented by a light blue colour and −1 spins are represented by a dark blue colour. Each spin has energy, which is defined as a negative summation of products of the positive spin with each of its four neighbouring spins. If there exist four opposing spins around a spin, then the energy is 4-positive, which is the maximum possible. However, the minimum possible energy, which is 4-negative, occurs when the spin is surrounded by four liking spins. Here, the energy is a measure of how many similar or opposite neighbours the spin has. A spin decides whether to "flip" to its opposite in the following way. For the spins, a low-energy state is favourable thus, a spin always tends to flip if flipping results decrease its energy. However, at times, the spins also flip into a higher-energy state.

The probability of flipping can be calculated through the Metropolis algorithm, which works as follows.

- Flipping probability: e-Ediff/temperature, where Ediff is the potential gain in energy.
- With an increase in the temperature, the probability of flipping to a higher-energy state becomes more likely; however, as the energy to be gained through flipping increases the probability of flipping decreases.

- To run the model, we recurrently pick a single random spin and give it the chance to flip.

Now, let us try correlate the Ising model with opinion formation by assuming that any person (spin) can take just two values, "yes" or "no" (up or down) and temperature is analogous to external information influencing opinion formation.

Figure 1a depicts that if flipping up or down, the probability is equal and the temperature is 0, then a state is reached where spins are equally aligned up and down. Comparing the same agents with evolving opinions, where no external agency is present to influence opinion formation, we find that with time, a state is reached in a population where an equally probable proportion of opinions exists. Figure 1b shows that increasing the temperature to 2 favours a spin-up state; this is similar to the scenario, whereby the existence of an external agency propagating some information shifts the opinions to either side, out of two available options. However, increasing the temperature to 5 depicts a more chaotic scenario in the case of spins, which is analogous to the prominent presence of some information-propagating agency. In this scenario, a state of mixed opinions is witnessed. Despite its simplistic and interesting approach, the Ising model is very simple to actually account for the complexity of each individual position and of individuals' interactions.

We correlated a grain growth phenomenon occurring at a microscopic level with real-world opinion formation and have performed computer simulation using NetLogo.

4   Grain Growth Phenomenon Simulations

Grain growth is an extensively investigated phenomenon in materials science: bigger grains grow at the expense of small grains (the overall volume is maintained). In the real world, opinion formation exhibits the similar phenomenon of the influence of the majority opinion over the minority one.

The agent-based simulation of grain evolution considered here is based on the principle of thermodynamics of atomic interactions. Initially, the material is represented as a hexagonal matrix or a square, in which each site resembles an atom and is assigned a value representative of its crystallographic orientation. The same orientation sections (contiguous regions) represent the grains. The grain boundaries are fictional surfaces separating the volumes with different orientations. The simulation is based on a simple basic algorithm: each atom continuously tries to be as stable as possible. The stability of atom is facilitated by the number of neighbours with similar orientations: the more similar the neighbours, the more stable it is. The presence of a few similar neighbours makes the atom to a more stable position.

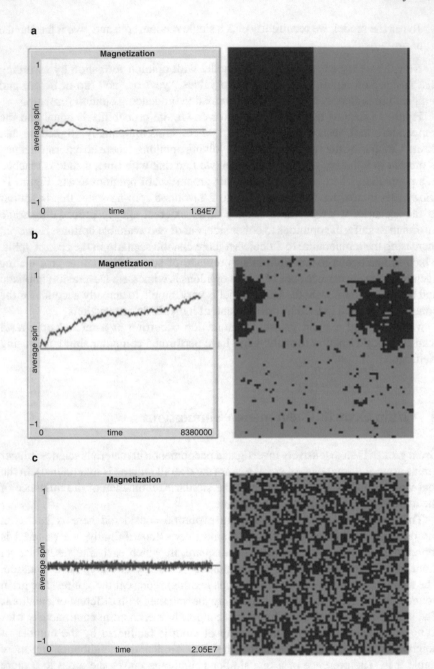

Fig. 1 Simulations of the Ising model using NetLogo with the probability of spin of up to 50% at temperature (**a**) 0, (**b**) 2 and (**c**) 5

An external factor such as annealing temperature is also a crucial factor in achieving the stable state. A high annealing temperature makes some atoms randomly jump to the next grain. When it is zero, only energetically favourable jumps will happen. Temperature here is represented as a relative value: a percentage of random jumps out of the possible jumps (100% means that all the jumps are random; Fig. 2).

- For 0% annealing temperature, there is no external energy to the atoms so they attain a stable state naturally by encountering the neighbouring atoms of the same orientation. Similarly, in the absence of any external agency, in the on-line social network, the members interact with neighbours of a similar nature and a majority stable state of opinion is reached naturally.
- For 5% annealing temperature the atoms become perturbed and jump to the next grain; this delays the achievement of the stable state. Likewise, even the small presence of an external information propagating agency starts to delay opinion formation.
- An increase in annealing temperature to 10 and 20% further increases the perturbation due to which the atom tends to make more random jumps to grains. This further delays the time to achieve the stable state compared with 0 and 5% annealing temperature. A similar phenomenon is observed in an on-line social network when the presence of any strong external agency influences the individual's opinion formation to such an extent that the presence of immediate neighbours becomes ineffective.
- Interestingly, for 100% annealing temperature, the attainment of a stable state is relatively fast compared with the case of 10 and 20% annealing temperature. This is similar to a situation where an online social network is under the presence of a very strong external agency, which accelerates opinion formation.

In this section, we have correlated the grain boundary crossing behaviour of atoms with the dynamics of opinion formation in an online social network.

In the next section, we used a demographic problem based on the Ising model to develop a model of opinion dynamics. Here, we considered two online communities with variable growth rates. We then examined the probability of either of the two populations or respective opinions dominating at different times, taking into account the probabilities of mixing, the new member joining rate etc.

5 Model Based on Population Dynamics

We consider two communities denoted by A and B. Community A holds opinion 1 and community B holds opinion -1. Let community A be larger than community B. The system under consideration is composed of N nodes; hence, each community has N/2 nodes. Every node is occupied by one agent i with opinion $O_i = 1$ (agree) or $O_i = -1$ (disagree), and the link between the two nodes in the same

Fig. 2 Simulations of grain growth using NetLogo at different annealing temperatures. (a) Annealing temperature 0% (b) Annealing temperature 5% (c) Annealing temperature 10% (d) Annealing temperature 20% (e) Annealing temperature 100%

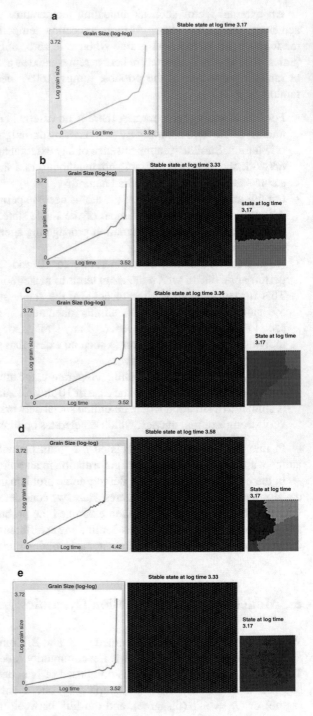

community represents the frequency of contact between them. Let community A, be characterized by its number $n_a(t)$, where t is the instantaneous time and the new member joining rate is a_n. The community B is characterized by $n_b(t)$ and the new member joining rate is b_n.

Our model is described by the following parameters:

1. The initial population of the two communities A and B in numbers are represented as $n_a(0)$, $n_b(0)$ respectively and their ratio:

$$R = \frac{n_a(0)}{n_b(0)} \tag{1}$$

2. The individual joining rates (or inflow into them) of the communities A and B are a_n and b_n respectively and their ratio:

$$J = \frac{a_n}{b_n} \tag{2}$$

3. The rate of joining by a new member in either of the communities depends on the total number of individuals in that community, for example, for community B it is:

$$b_n^{(1)} = b_n + (a_n - b_n) \times (1 - V(t)) \tag{3}$$

Where $V_i(t) = 1 - J \times \frac{n_a(t) + n_b(t)}{N}$.

4. The switch-over rate of opinion m is equally probable in both communities,

$$dn_a(t) = a_n n_a(t)dt - mn_a(t)dt + mn_b(t)dt$$

$$= a_n n_a(t)dt \left(1 - \frac{m}{a_n} \left(\frac{n_a(t) - n_b(t)}{n_a(t)} \right) \right) \tag{4}$$

$$dn_b(t) = b_n n_b(t)dt - mn_b(t)dt + mn_a(t)dt$$

$$= b_n n_b(t)dt \left(1 - \frac{m}{b_n} \left(\frac{n_b(t) - n_a(t)}{n_b(t)} \right) \right) \tag{5}$$

5. The switch-over rate of opinions from community A to community B is signified by m and that from community B to community A is signified by m':

$$dn_a(t) = a_n n_a(t)dt - mn_a(t)dt + m'n_b(t)dt$$

$$= a_n n_a(t)dt \left(1 - \frac{m}{a_n} \left(1 - \frac{m'n_b}{mn_a} \right) \right) \tag{6}$$

$$dn_b(t) = b_n n_b(t)dt - mn_b(t)dt + m'n_a(t)dt$$

$$= b_n n_b(t)dt \left(1 - \frac{m}{b_n} \left(1 - \frac{mn_a}{m'n_b} \right) \right) \tag{7}$$

6. The parameters required for the analysis of opinions are:

$$r = \frac{m}{a_n}$$

$$\frac{m}{b_n} = J \times \frac{m}{a_n}$$

$$p = \frac{n_a}{n_b}$$

$$\alpha = \frac{m}{m'}$$

7. The final equations of the rate of joining with the mixing parameter, m, for both the communities are:

$$\frac{dn_a(t)}{dt} = a_n n_a(t) \left(1 - r \times \left(1 - \frac{1}{p}\right)\right) \tag{8}$$

$$\frac{dn_b(t)}{dt} = b_n n_b(t)(1 - J \times r \times (1 - p)) \tag{9}$$

The final equations of the rate of joining with mixing parameters, m and m' are:

$$\frac{dn_a(t)}{dt} = a_n n_a(t) \left(1 - r \times \left(1 - \frac{1}{\alpha p}\right)\right) \tag{10}$$

$$\frac{dn_b(t)}{dt} = b_n n_b(t) \left(1 - J \times \frac{r}{\alpha} \times (1 - p\alpha)\right) \tag{11}$$

Assume that the intercommunity interaction is absent and hence the two communities do not influence the opinions of each other as shown by Fig. 3. However, the factor J now plays a crucial role. If the rate of new members joining community B is very high compared with joining community A, there exists a disordered state. Hence, the densities of the two opinions are equal and no opinion dominates.

Now consider the case in which the rates of joining community A (a_n) and community B (b_n) are the same, as shown by Fig. 4. The probability of an individual from community A migrating to community B (m) is also taken to be same in this scenario. We observe that with time, the population of community A increases rapidly in the case when no migration ($m = 0$) from A to B takes place. The initial population of A is higher; thus, the individuals migrating from A to B ($m \times n_a$) are more than individually migrating from B to A ($m' \times n_b$). Therefore, the gain in population takes place over time if migration does not occur.

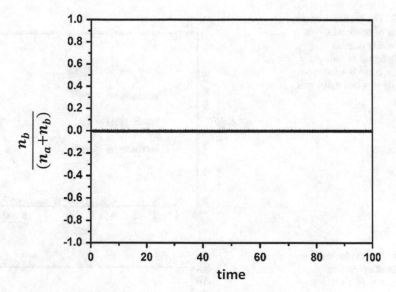

Fig. 3 Case of absence of intercommunity interactions

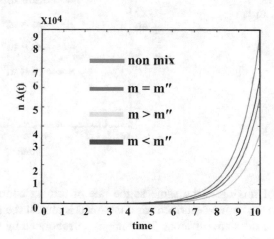

Fig. 4 The evolution of the growth of opinion in community A, with parameters is as follows: $a_n = 1, b_n = 1, m = 1, n_a(0) = 4$ and $n_b(0) = 1$

Comparing this with opinion dynamics, in this case, the interchange of opinion (termed as mixing) is not allowed i.e. an individual cannot change his/her opinion, the opinion of A dominates, but as the size of increase of opinions is the same, opinion A is always greater than opinion B. If mixing is allowed, the opinion of B increases as the change of opinion from A to B ($m \times N_a$) is greater than that from B to A ($m' \times N_b$) (Fig. 5).

For the parameters: $a_n = 1, b_n = 0.4, m = 1, n_a(0) = 4$ and $n_b(0) = 1$, the rate of joining community A is higher than the rate of joining community B and

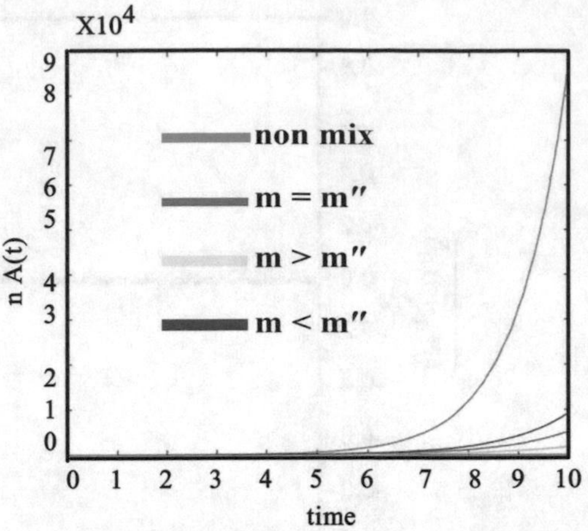

Fig. 5 The evolution of the growth of opinion in community A, with parameters is as follows: $a_n = 1, b_n = 0.4, m = 1, n_a(0) = 4$ and $n_b(0) = 1$

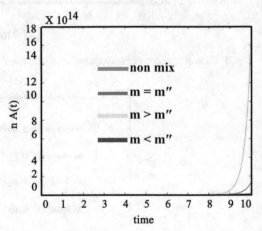

Fig. 6 The evolution of the growth of opinion in community A, with parameters is as follows: $a_n = 1, b_n = 4, m = 1, n_a(0) = 4$ and $n_b(0) = 1$

the probability of mixing is the same as the rate of joining community A. This case is similar to the above-described case; the trend is almost the same. However, the impacts of the rates are different. The same is corroborated by the evolution of opinions (Fig. 6).

With parameters such as $a_n = 1, b_n = 4, m = 1, n_a(0) = 4$ and $n_b(0) = 1$, the probability of joining community B is higher than that of joining community A and that of mixing is the same as the probability of joining community A. In this case, community A increases if there is an increase in community B. If the size of community B increases enormously, then the probability of members from B shifting to A increases. However, the joining rate remains constant. When the rate

Fig. 7 The evolution of the growth of opinion in community A, with parameters is as follows: $a_n = 1, b_n = 4, m = 0.2, n_a(0) = 4$ and $n_b(0) = 1$

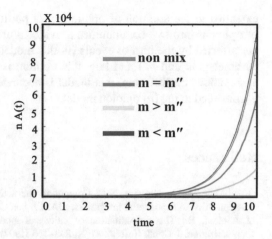

of mixing is higher or lower than the rate of joining community B, the same trend is followed as for the rate of joining community A. The impact thus observed is different and the curves (opinion of A) reach different values, but overall the trend remains the same, except for the case in which $m < a_n$ and $a_n = b_n$.

Here, the density of opinion of A will be less than that of B with evolution of time, as the number of people accepting opinion B is higher than those accepting opinion A. However, analysis depicts the opinions of A (minority opinion) still being at a maximum, facilitated by maximum probability of people changing their opinion from B to A (Fig. 7).

In this case, the maximal increase in the population of community A is a result of the higher probability of B mixing with A compared with A mixing with B.

From the above "two-population model," we are able to analyze the opinion dynamics of two opinions over time and see the opinion at a particular time. There may be more than two populations in a region; thus, we can extend the same arguments and analyze the opinion dynamics of a system in which there are a variety of opinions.

6 Conclusion

We have demonstrated the models concerning opinion dynamics that incorporated the simulations based on the grain growth phenomenon of atoms and also the two population growth dynamics. Grain growth simulations clearly indicated how the presence of external factors such as temperature in the case of atoms and mass media in the case of online communities affect the process of grain growth in the case of atoms and opinion formation in the case of online communities. The proposed model based on population dynamics establishes a firm confirmation of the fact that it is compatible with opinion dynamics. Also, the concept of birth rate is correctly extended to the addition of opinions in a particular community. The segregation of opinions into two communities provided a brilliant insight into how the masses are affected by the turn of events (in this model, the events are well replicated by parameters m and m') and how this opinion is dictated by the factors of inflow and outflow. Furthermore, this model is accurate when the binary opinion system is converted to the (n) opinion model.

References

1. Al-Mohy, A.H., Higham, N.J.: Computing the action of the matrix exponential, with an application to exponential integrators. SIAM J. Sci. Comput. **33**(2), 488–511 (2011)
2. Axelrod, R.: The dissemination of culture a model with local convergence and global polarization. J. Confl. Resolut. **41**(2), 203–226 (1997)
3. Bass, F.M.: A new product growth for model consumer durables. Manag. Sci. **15**(5), 215–227 (1969)
4. Benczik, I.J., Benczik, S.Z., Schmittmann, B., Zia, R.K.P.: Opinion dynamics on an adaptive random network. Phys. Rev. E **79**(4), 046104 (2009)
5. Birgin, E.G., Martínez, J.M., Raydan, M.: Nonmonotone spectral projected gradient methods on convex sets. SIAM J. Optim. **10**(4), 1196–1211 (2000)
6. Caceres, C.H., Wilkinson, D.S.: Large strain behaviour of a superplastic copper alloy-I. Deformation. Acta Metall. **32**(3), 415–422 (1984)
7. Castelló, X., Baronchelli, A., Loreto, V.: Consensus and ordering in language dynamics. Eur. Phys. J. B **71**(4), 557–564 (2009)
8. Clark, M.A., Alden, T.H.: Deformation enhanced grain growth in a superplastic Sn-1% Bi alloy. Acta Metall. **21**(9), 1195–1206 (1973)
9. Clifford, P., Sudbury, A.: A model for spatial conflict. Biometrika **60**(3), 581–588 (1973)
10. Cohen, R., Erez, K., Ben-Avraham, D., Havlin, S.: Resilience of the Internet to random breakdowns. Phys. Rev. Lett. **85**(21), 4626 (2000)
11. Cox, J.: Coalescing random walks and voter model consensus times on the torus in Zd. Ann. Probab. **17**(4), 1333 (1989)
12. Das, A., Gollapudi, S., Munagala, K.: Modelling opinion dynamics in social networks. In: Proceedings of the 7th ACM International Conference on Web Search and Data Mining, pp. 403–412. ACM, New York (2014)
13. De, A., Bhattacharya, S., Bhattacharya, P., Ganguly, N., Chakrabarti, S.: Learning a linear influence model from transient opinion dynamics. In: Proceedings of the 23rd ACM International Conference on Conference on Information and Knowledge Management, pp. 401–410. ACM, New York (2014)
14. Deffuant, G., Neau, D., Amblard, F., Weisbuch, G.: Mixing beliefs among interacting agents. Adv. Complex Syst. **3**(1–4), 87–98 (2000)
15. DeGroot, M.H.: Reaching a consensus. J. Am. Stat. Assoc. **69**(345), 118–121 (1974)
16. Dornic, I., Chaté, H., Chave, J., Hinrichsen, H.: Critical coarsening without surface tension: the universality class of the voter model. Phys. Rev. Lett. **87**(4), 045701 (2001)
17. Erdös, P., R'enyi, A.: Statistical physics of social dynamics. Publ. Math. Debr. **6**(290) (1959)
18. Farajtabar, M., Du, N., Gomez-Rodriguez, M., Valera, I., Zha, H., Song, L.: Shaping social activity by incentivizing users. In: Advances in Neural Information Processing Systems, pp. 2474–2482 (2014)
19. Fernández-Gracia, J., Eguíluz, V.M., San Miguel, M.: Update rules and interevent time distributions: slow ordering versus no ordering in the voter model. Phys. Rev. E **84**(1), 015103 (2011)

20. Friedkin, N.E., Johnsen, E.C.: Social influence and opinions. J. Math. Sociol. **15**(3–4), 193–206 (1990)
21. Galam, S.: Minority opinion spreading in random geometry. Eur. Phys. J. B **25**(4), 403–406 (2002)
22. Granovetter, M.: Threshold models of collective behavior. Am. J. Sociol. 1420–1443 (1978)
23. Hegselmann, R., Krause, U.: Opinion dynamics and bounded confidence models, analysis, and simulation. J. Artif. Soc. Soc. Simulat. **5**(3) (2002)
24. Hegselmann, R., Krause, U.: Opinion dynamics driven by various ways of averaging. Comput. Econ. **25**(4), 381–405 (2005)
25. Herriot, G., Suery, M., Baudelet, B.: Superplastic behaviour of the industrial Cu7wt.% P alloy. Scripta Metall. **6**, 657 (1972)
26. Holley, R.A., Liggett, T.M.: Ergodic theorems for weakly interacting infinite systems and the voter model. Ann. Probab. **3**, 643–663 (1975)
27. Holme, P.: Modern temporal network theory: a colloquium. Eur. Phys. J. B **88**(9), 1–30 (2015)
28. Kaur, R., Kumar, R., Bhondekar, A.P., Kapur, P.: Human opinion dynamics: an inspiration to solve complex optimization problems. Sci. Rep. **3**, 3008 (2013)
29. Krause, S.M., Bornholdt, S.: Opinion formation model for markets with a social temperature and fear. Phys. Rev. E **86**(5), 056106 (2012)
30. Leone, M., Vázquez, A., Vespignani, A., Zecchina, R.: Ferromagnetic ordering in graphs with arbitrary degree distribution. Eur. Phys. J. B **28**(2), 191–197 (2002)
31. Metropolis, N., Rosenbluth, A.W., Rosenbluth, M.N., Teller, A.H., Teller, E.: Equation of state calculations by fast computing machines. J. Chem. Phys. **21**(6), 1087–1092 (1953)
32. Mobilia, M.: J. Stat. Phys. **151**(1–2), 69 (2013)
33. Newman, M.E.: Assortative mixing in networks. Phys. Rev. Lett. **89**(20), 208701 (2002)
34. Pineda, M., Toral, R., Hernandez-Garcia, E.: Noisy continuous-opinion dynamics. J. Stat. Mech Theory Exp. **2009**(08), P08001 (2009)
35. Schweitzer, F., Behera, L.: Nonlinear voter models: the transition from invasion to coexistence. Eur. Phys. J B **67**(3), 301 318 (2009)
36. Shao, J., Havlin, S., Stanley, H.E.: Dynamic opinion model and invasion percolation. Phys. Rev. Lett. **103**(1), 018701 (2009)
37. Stauffer, D.: Sociophysics: the Sznajd model and its applications. Comput. Phys. Commun. **146**(1), 93–98 (2002)
38. Suery, M., Baudelet, B.: Flow stress and microstructure in superplastic 60/40 brass. J. Mater. Sci. **8**(3), 363–369 (1973)
39. Takaguchi, T., Masuda, N.: Voter model with non-Poissonian interevent intervals. Phys. Rev. E **84**(3), 036115 (2011)
40. Volovik, D., Redner, S.: Dynamics of confident voting. J. Stat. Mech Theory Exp. **2012**(04), P04003 (2012)
41. Watts, B.M., Stowell, M.J., Cottingham, D.M.: The variation in flow stress and microstructure during superplastic deformation of the Al-Cu eutectic. J. Mater. Sci. **6**(3), 228–237 (1971)
42. Wilensky, U., Stroup, W.: HubNet (1999). http://ccl.northwestern.edu/netlogo/hubnet.html. Center for Connected Learning and Computer-Based Modelling, Northwestern University. Evanston, IL
43. Zhu, C. P., Kong, H., Li, L., Gu, Z. M., & Xiong, S. J.: An inverse voter model for co-evolutionary networks: Stationary efficiency and phase transitions. Phys. Lett. A. **375**(11), 1378–1384 (2011)
44. Zschaler, G., Böhme, G.A., Seißinger, M., Huepe, C., Gross, T.: Early fragmentation in the adaptive voter model on directed networks. Phys. Rev. E **85**(4), 046107 (2012)

Opinion Mining from Social Travel Networks

Charu Puri, Akhil Yadav, Gaurav Jangra, Kush Saini, and Naveen Kumar

Abstract Opinions have always been an important part of daily lives as it has been a human behaviour to gather information about what other people think about the things we want to buy or the hotel we want to stay at or the place we want to visit. These opinions, whether positive or negative, are backed up by the opinion of many other people, which enables us to start to build up our own opinion about that thing or place. Over time, the number of information-gathering sources have increased, new opportunities and challenges have arisen, and people have been actively providing their opinions on different websites and using information-gathering technologies to seek out and understand the sentiment and opinions. Opinion mining deals with the extraction of emotions expressed in natural language using a digital form. It is being used in almost every field to monitor the emotions and sentiments of people. The purpose of this chapter is to study and analyze the emotions expressed by people about their travel experiences.

Keywords Lexeme • Opinion mining • Polarity • Scrapy • Social network • Supervised learning • Travel website

1 Introduction

A social network is a medium for friends, family, acquaintances, colleagues and strangers to interact, give views and to exchange ideas, facts and opinions. With a surge of travellers accessing the websites, the current trend has shifted from accessing the sites, not for mere travel information, rather travellers are now making use of social media more than ever to help influence their purchasing decisions for a trip. People research extensively before booking a hotel and turn to social media for their answers to get opinions from other travellers. For example, initially, TripAdvisor.com worked as a travel website rather than as a social networking site. Over time, it started assisting travellers by providing travel information, reviews

C. Puri (✉) • A. Yadav • G. Jangra • K. Saini • N. Kumar
Department of Computer Science, University of Delhi, New Delhi, India
e-mail: cpuri.cs.du@gmail.com; akhilgm2@gmail.com; gjangra4u@gmail.com; kushsaini.700@gmail.com; nk.cs.du@gmail.com

© Springer International Publishing AG 2017
H. Banati et al. (eds.), *Hybrid Intelligence for Social Networks*,
DOI 10.1007/978-3-319-65139-2_8

of travel-related content and by forming a user-interactive model that somewhat resembles a social networking website. Travellers not only seek information from travel sites, they also post their opinions in the form of audio, video, images and text. The opinions are often expressed using terms such as excellent, very good, good, bad, very bad by using star ratings.

Data mining is the process of analyzing data and brings out the useful information—information that can be used to increase productivity, revenue, reduce costs, search for information or make decisions [2].

Determining the customer's opinion about a newly launched product is of utmost importance for determining the directions of development. Opinion mining is a discipline that deals with automated detection and extraction of opinions and sentiments. Opinion mining applications provide efficient referring systems, statistical analysis, market trends and product development [17]. Figure 1 explains the flow from data mining to opinion mining. As Fig. 1 shows, data mining, which deals with consumer behaviour, reviews or their sentiments, is studied under opinion mining, also known as sentiment analysis. Opinion mining is a part of data mining and an application of natural language processing (NLP). Research work in the field of opinion mining was mainly carried out only after the late 1990s and early 2000s, and because of the range of applications, industrial interest, social media and many other factors, it has progressed in this field. Owing to the real-life applications of opinion mining, it has become a popular research problem and has been worked upon. It is extensively analyzed in data science, Web mining and information retrieval systems. It has spread into multiple disciplines, whether computer science or management sciences, and it is used extensively everywhere.

The five major elements of Opinion Mining are [3]:

- Crawl, modify and push data onto the database system.
- Maintain the data in a database system.

Fig. 1 Opinion mining

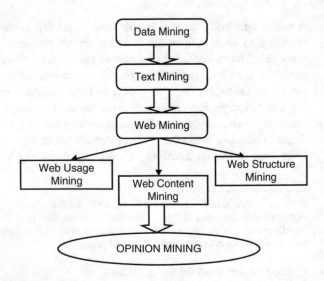

- Provide data to analysts and information technology professionals.
- Analyze the data using application software.
- Present the data in a useful and understandable format

The objective is to classify sentiments at a different level namely: the document level, the sentence level and the feature/entity level. Different classification levels have some limitations such as classification at:

The document level could only be done if the document has reviews for a single entity. In practice, the Q&A websites and blogs depend on sentiment classification to find users' opinions about various entities and topics.

Sentence level classification goes deeper than the document-level sentiment classification, as it relates closer to opinion targets and sentiments on the targets. However, it also has some limitations such as:

- Complex sentences give different sentiments on targets depending upon the situation
- Although a sentence may have an overall positive or negative tone, some of its components may express opposite opinions
- Sentence level classification faces challenges in dealing with comparative sentences

Aspect-level sentiment analysis is then used for real-life purposes because of the level of detail. Most industrial systems are also based on aspect level classification. Although many efforts have been made in the research community regarding this problem, and many systems have been built, but it is still far from being solved. Every sub-problem remains highly challenging. The two basic problems of entity level classification are aspect extraction and aspect sentiment classification.

2 Opinion Mining

Opinion mining, also called sentiment analysis, is the discipline that analyzes people's opinions, sentiments, attitudes, appraisals, evaluations and emotions towards entities such as products, sales services, organisation, issues, events and their attributes [7]. It has lots of problems space and goes by different names for different tasks, such as sentiment analysis, opinion mining, opinion extraction, sentiment mining, subjectivity analysis, emotion analysis and review mining. In the paper [8], the authors have described the need to assess the added value of information available before and after the post's creation time in the sentiment analysis of Facebook posts; determine the priority of predictors; and investigate the relationship between predictors and sentiment. The paper [10] gives the scientometric analysis of research work done in the field of sentiment analysis from 2000 to 2015 using research publications from Web of Science (WoS) as a dataset. In paper [4], the author gives a fast, flexible and generic methodology for sentiment detection out of textual snippets that express people's opinions in different languages.

2.1 Sentiment Analysis and Classification

Sentiment classification is a process of classifying the text as per the polarity, whether positive or negative, of the sentiment expressed in the opinion. Online reviews have rating values assigned by the reviewers and the assigned ratings determine the polarity of review, e.g. 1–5 stars. A review with a 4- or 5-star rating has a positive polarity, whereas a review with 1–2 stars is considered to be a review with negative polarity. A neutral class can also be considered if reviews are assigned 3 stars. Sentiment classification is a text classification problem in which text is classified based on different topics of the document, e.g. spiritual, science and sports. Words related to the topic are key features and are very relevant in such classifications. In sentiment classification, words that highlight positive or negative sentiments are important, for example, bravo, marvellous, amazing, great, gorgeous, horrible, drastic, blunder and worsened [7]. Sentiments can be analyzed at three levels as:

- Document level: In this case, the task is to analyze the polarity of the sentiment document. For example, analyzing the provided product review, the system determines the positivity or negativity, i.e. the polarity of the opinion given of the product. This classification is known as document-level sentiment classification. In document-level classification, the sentiment document must have reviews about a single entity only. It is not capable of processing documents that assess different entities.

Definition 2.1 Problem definition: Provided a sentiment document "d" assessing the given entity, objective is to decide the comprehensive opinion "s" of the opinion holder about the entity, i.e., determine "s" expressed as the aspect GENERAL in the quintuple $(\_, GENERAL, s, \_, \_)$, where the entity "e", opinion holder "h", and time of opinion "t" are considered insignificant or known. Now, "s" can take two types of value: categorical value, positive or negative, and numerical value, 1 to 5 or any other defined value. If "s" is assigned a categorical value, it classifies as a classification problem, and if "s" is assigned some value within a given range, it classifies as a regression problem [6].

Document classification is hard to apply to non-review sentences such as in discussion forums, online blogs and news articles, because such posts usually evaluate multiple products for comparison. In these cases, it is very hard to determine whether the post even has any sentiment or opinion expressed, if a sentiment is expressed or just a fact. Sentiment classification on a document level does not obtain high-quality results, as in-depth natural language processing is required for better results and is not performed at this level of classification. Online reviews usually have a rating, which is also given by the same opinion holder that is posting the review.

- Sentence level: Document-level classification too broad for most applications, which raised the need to move into the depth of sentence-level classification, i.e. classifying opinions expressed in sentences. Nonetheless, the difference between

"We are very satisfied."
⊙⊙⊙⊙○ Reviewed 19 January 2011

gertenhong
Utrecht, The Netherlands
Ⓐ 1 review
Ⓓ 1 helpful vote

Outside our room there was a mini-bar. There we eat our breakfast and there is a watercooker and other supplies. You can rent DVD at the hotel office.My wife watch every day "Holliday in Rome".Close to the hotel there is a small bar and restaraunt with nice employees and delicious food.

Helpful?   👍 1   Thank gertenhong   ⚑ Report

Fig. 2 Example

the document- and sentence-level classification is very narrow, as sentences themselves are nothing but smaller documents. A sentence usually consists of only one sentiment, but this may not always be true, whereas a document typically contains multiple opinions. Starting with an example in Fig. 2.

The first sentence in Fig. 2 states a fact and expresses no sentiment, it is classified as an objective sentence. Other sentences show sentiments in either an explicit or an implicit way.

The objective at this level is to determine the polarity of the sentence. Polarity can be positive, negative or neutral. Sentence-level classification determines the subjectivity of the sentence, which distinguishes sentences that express some fact in the given information from sentences that express subjective views and opinions, i.e. objective and subjective sentences.

Definition 2.2 Problem definition: Given a sentence x, determine the polarity of x and classify them as a positive or negative opinion, or a neutral opinion when needed. The quintuple (e, a, s, h, t)—where "e" is the entity, "an" is the aspect of the entity, "s" is the sentiment expressed, "h" is the opinion holder and "t" is the time of the opinion.

This definition is not used at a sentence level as this classification is an intermediate step. For sentence-level classification, the opinion targets are to be identified. If the polarity of the opinion is known, but not what entities/aspects the opinion is about, then it is of limited use. However, if the entity or aspect of the entity is known, then sentence-level classification can be used and polarity can be determined [7].

Definition 2.3 Objective sentences state some fact or straightforward information that expresses no sentiment or opinion, with implicit sentences as an exception.

The review shown in Fig. 3 is an objective paragraph. It implies a positive sentiment about the reachability because of the described fact. Thus, it is better to classify each sentence as opinionated or not opinionated, regardless of whether it is subjective or objective. Existing research on sentence-level subjectivity

Fig. 3 Example

classification is primarily focused on solving the problem without considering that different treatments are needed for different types of sentences.

Definition 2.4 Conditional sentences describe notional or hypothetical situations and their consequences. Such a sentence typically contains two clauses: the condition clause and the consequent clause, which are dependent on each other. This relationship has a significant impact on the polarity of a sentence.

A point to be noted is that sentiment words alone cannot distinguish a subjective sentence from an objective one, e.g. *west coast custom deals in modified cars, pay them a visit and "Nutella tastes better than Hershey's syrup!"*. The first sentence expresses no sentiment towards any product, but the second sentence leans positively towards Nutella and compares Nutella with Hershey's syrup. Clearly, non-conditional sentences differ from conditional sentences and hence have different sentiment analysis methods.

There are four types of sentences: interrogative sentences, sarcastic sentences, exclamatory sentences and comparative sentences. Considering a comparative sentence, sentiments in comparative sentences cannot be classified at a sentence level. For example, *Rome is better than Beijing*. Here, different methods are used to extract and analyze comparative opinions, as comparative sentences have different meanings from regular opinions. Although this sentence clearly expresses an opinion, determining its polarity is not simple.

- Entity and aspect level: As the classification at a document level and sentence level cannot evaluate the sentiments, aspect-level classification performs in-depth analysis. Instead of looking at language constructs, aspect classification focuses directly on the sentiment expressed. It is based on the idea that a sentence consists of a sentiment with some polarity, and a target of opinion. Now, clustering opinion reviews of a single entity with all positive or all negative opinions does not mean that all entities have positive or negative opinions. For detailed analysis, the aspects are determined that give the polarity of the sentiment for each aspect [7].

At the aspect level, the objective is to discover every quintuple $(e_i, a_{ij}, s_{ijkl}, h_k, t_l)$ in a given document d and to complete this, two core tasks are aspect sentiment classification and aspect extraction.

- Aspect sentiment classification: Here, the orientation of the expressed sentiment on each aspect of an entity is determined. There are two main approaches:

  - The supervised learning approach, and
  - The lexicon-based approach.

Supervised learning is dependent on the training data. It is a model or classifier, as discussed at the document level, trained from labelled data in one domain that often performs poorly in another domain. The lexicon-based approach can avoid some of the issues, and performs quite well in multiple domains. These methods are usually called unsupervised learning, which uses a sentiment lexicon consisting of a list of sentiment words, composite expressions, rules of opinions to assess the orientation of a sentiment on a given aspect in the sentence. An opinion rule expresses a concept that implies a positive or negative sentiment. Sometimes, it is as simple as an individual sentiment word with implied sentiments or as compound expressions that may need deeper knowledge to determine the orientations.

- Aspect extraction: The process of extracting information from the entity is known as aspect extraction. In sentiment mining, the main task is to determine the target, which is often the aspect of an entity that is to be extracted. This makes it important to find out each opinion expression and its target from the sentence. Some opinion expressions can also play dual roles as expressions can act as the aspect and express a sentiment too [9]. Here, the focus should be on the aspect extraction task with four main approaches [7]:

  - Extraction based upon nouns and noun phrases: In this method, aspect expressions from noun and noun phrases from multiple reviews in each domain are highlighted. Out of the domain, the nouns are determined using a part-of-speech tagger. The most frequent nouns are counted and the most frequently occurring ones are retained. The reason why the most frequent words are retained is because when people comment upon a single entity, their vocabulary converges. Thus, the frequent ones are considered as the aspect. There are many algorithms for finding the aspect, one of which works in a such way that it evaluates each discovered noun phrase by computing a pointwise mutual information (PMI) score between the phrase and metonymy discriminators associated with the entity class. For example, if a camera is the target and some of the camera reviews are considered, then the metonymy discriminator may be *camera was, camera has, camera comes with, of camera, etc.*

Then the *PMI* measure is given by

$$\mathrm{PMI}(a, d) = (\mathrm{hits}(ad))/(\mathrm{hits}(a)\mathrm{hits}(d)),$$

where a is a candidate aspect determined using the frequency approach and "d" is a discriminator. Here, the idea of this approach is clear. As a and d do not occur frequently, this results in a low value of *PMI* showing that aspect is not a part of the considered product.

– Using opinion and target relations: As opinions have targets, opinions are obviously related. This relationship is exploited to extract aspects that are opinion targets because sentiment words are often known. The idea is as follows. A single sentiment word is used to describe or modify different aspects. If a sentence does not have a frequent aspect, but has some sentiment words, the nearest noun or noun phrase to each sentiment word is extracted. For example, *in this software is amazing*, since the sentiment word *amazing* is known, and the nearest noun is *software*, it is extracted as an aspect.

– Using supervised learning Aspect extraction is special case of a general information extraction problem. The most important method of supervised learning is sequential learning. As these are supervised techniques, manually labelled aspects are required in the text body.

Based upon the above-mentioned data, the example described below can be easily understood.

Example "It was a wonderful stay at this hotel. Our host, Emilio, was very helpful and accommodating. Our rooms were the best part, they were large and airy with a beautiful view from the window. We were surprised when we saw the towel heating element in the bathroom. I bathrooms were sleek and modern. There was plenty of closet space and shelves for all our clothes. We would love to stay here again whenever we return to Rome."

Depending upon the condition, this review can be seen at the: Document level, i.e. is this review + or −? Sentence level, i.e. is each sentence + or −? Entity and feature/aspect level. A regular opinion or a comparative opinion is shown in Fig. 4. Sentiment analysis in this respect can be considered, as in Fig. 4, an opinion classification process. The target of sentiment analysis is to find opinions, identify the sentiments expressed, and then classify their polarity.

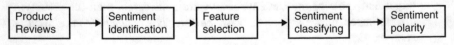

Fig. 4 Example

2.2 Type of Opinions

While writing in natural language, opinions are written in ways that may give a direct opinion, may state a fact or may give a comparison between entities. There are many types and subtypes of opinions that are stated as:

- Regular opinion: A regular opinion is just a simple sentiment expressed with two sub-types.
- Direct opinion: A direct opinion expresses sentiment on an entity or aspect, e.g. the sentence in Fig. 5.

    Great value for money.
- Indirect opinion: An indirect opinion is an opinion that is expressed indirectly on an entity or aspect based on its effects on other entities. For example, the sentence in Fig. 5 *Staff were there when I had any problems and thought of me as a friend* describes the friendly and caring nature of the hotel staff, which indirectly gives a positive opinion or sentiment to the hotel staff (Fig. 6).

    According to a recent trend, most research is carried out on direct opinions, which are simpler to handle than indirect opinions.
- Comparative opinion: A comparative opinion expresses a relation of similarities or differences between two or more entities. It also shows a preference of the opinion holder based on some shared aspects of the entities. For example, the sentences *Usain Bolt is faster than Gary Dunn and Vijender Singh is the best boxer* express two comparative opinions.

Fig. 5 Example

"Beautiful"
Reviewed 2 July 2016 via mobile
Staff was there when I had any trouble and they thought of me as a friend.It is at a Walking distance to the attractions, restaurants and a very short walk to the metro station. Highly recommended and ask for a view looking onto the opera house, stunning!!!! 100% would stay here again
catherineldx
Level 3 Contributor
11 reviews
4 hotel reviews
5 helpful votes
Helpful?
Thank catherineldx
Report

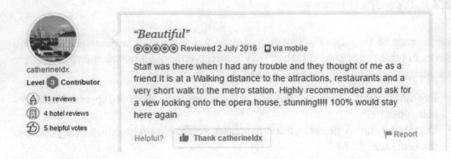

Fig. 6 Example

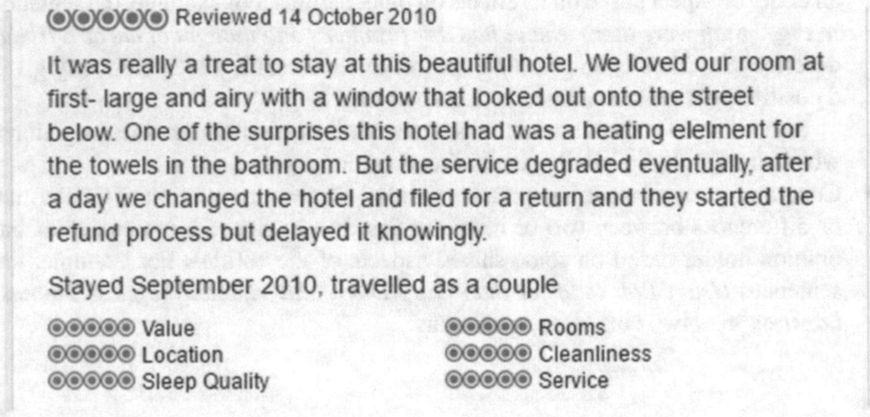

Reviewed 14 October 2010
It was really a treat to stay at this beautiful hotel. We loved our room at first- large and airy with a window that looked out onto the street below. One of the surprises this hotel had was a heating elelment for the towels in the bathroom. But the service degraded eventually, after a day we changed the hotel and filed for a return and they started the refund process but delayed it knowingly.
Stayed September 2010, travelled as a couple
Value
Location
Sleep Quality
Rooms
Cleanliness
Service

Fig. 7 Example

- Explicit opinion: An explicit opinion is a subjective statement that gives a regular or comparative opinion, *e.g. "Las Vegas has best casinos", and "Las Vegas has better casinos than Beijing"*.
- Implicit (or implied) opinion: An implicit opinion is an objective statement that implies a regular or comparative opinion. Such an objective statement usually expresses a desirable or undesirable fact, for example in Fig. 7.

Here, in Fig. 7, *the service degraded eventually, after a day we changed the hotel and filed for a return and they started the refund process but delayed it knowingly* is an example of implicit opinion.

Explicit opinions are easier to detect and to classify than implicit opinions, although much of the research work is focused on explicit opinions.

2.3 Sentiment Word and Its Issues

There are words that are commonly used to express positive and negative opinions. For example: adorable, robust and delightful are words with positive polarity, and bad, hardly and terrible are words with negative polarity. Apart from words, there are also phrases and idioms, e.g. cost someone an arm and a leg. A list of such words and phrases is called a sentiment lexicon. Although sentiment words and phrases are important, they are not sufficient, because:

- A word can have different meanings in different domains.
- A sentence containing sentiment words such as an interrogative sentence, may not express any sentiment.
- Sarcastic sentences with or without sentiment words are hard to deal with.
- Many sentences without sentiment words can also imply opinions.

For example:

- (1) We toured Italy for our honeymoon and the first stop was in Rome.
- (2) I simply love it.
- (3) The scenery from the room was amazing.
- (4) The refrigerator in the room was small
- (5) However, the stay was wonderful.

In the given five sentences, it can be seen that

- Sentence (1) does not express any opinion.
- Sentence (2) clearly states the opinion in an opinion word love.
- Sentence (3) also provides a clear view on the sentiment with respect to entity sensors quality with sentiment amazing.
- Sentence (4) is clearly an interrogative sentence without an opinion.
- Sentence (5) also clearly expresses the sentiment.

2.4 Subjectivity and Emotion

Subjectivity and emotion are two ideas that are closely related to sentiment and opinion.

Definition 2.5 Sentence subjectivity: An objective sentence presents some factual information about the world, whereas a subjective sentence expresses personal feelings, views, or beliefs.

- Objective sentence: An objective sentence presents some fact as information.

  - Objective sentences can imply opinions or sentiments because of desirable and undesirable facts.

Fig. 8 Example

Fig. 9 Example

- For example, the review in Fig. 8 *We toured Italy for our honeymoon and the first stop was in Rome. We stayed here for 4 days and 3 nights. The hotel is within walking distance of the Colosseum, Ancient City, their subway system and many good restaurants and bars. The rooms were very cute and clean* [18].

- Subjective sentence: A subjective sentence presents personal feelings, views or beliefs. There are various subjective expressions, such as opinions, allegations, desires, beliefs and suspicions.

  A subjective sentence is not bound to express a sentiment. For example, the review in Fig. 9: *staff were very helpful and nice, would recommend to everyone! Enjoyed the stay.*

- Emotions: are also studied and analyzed in fields such as psychology, philosophy and sociology. Emotions play a large role in everyday life and manipulate the decision-making ability. There is no fixed set of emotions that can represent all the emotions expressed by using natural language, but usually people show six primary emotions, such as love, joy, surprise, anger, sadness and fear, which can further be sub-divided into secondary emotions. Each emotion has a different intensity depending upon the situation.

Definition 2.6 Emotions: are subjective feelings and thoughts. Emotions are closely related to sentiments, with the strength of a sentiment being directly connected to the intensity of expressed emotions [7].

Opinions are mostly evaluations of sentiments, which are of two types:

- Emotional evaluation: Such evaluations are from non-tangible and emotional responses to entities that go deep into people's states of mind. For example, the following sentences express emotional evaluations: *"I love my mother", "I am so angry with their service" and "Mustangs are the best cars ever built".*
- Rational evaluation: Evaluations from rational reasoning, tangible beliefs and functional attitudes. For example, the sentences in the review shown in Fig. 9 express rational evaluations: *"This is a hotel that occupies some of the floors in an older building", "The bathroom is small but very modern" and "I am happy with the stay".*

To implement these evaluations in practice, sentiment ratings can be benchmarked as emotional negative (−2), rational negative (−1), neutral (0), rational positive (+1) and emotional positive (+2). In practice, neutral sentences do not express any opinions or sentiments. The concepts of emotion and opinion are clearly not equivalent. With rational opinions without sentiments, *e.g. "The voice of this phone is clear", and many emotional sentences that clearly express no sentiment on anything, e.g. "I am so surprised to see you here". Moreover, some emotions may not have targets, but express only people's feelings instead, e.g. "I am so sad today".*

3 Opinion Mining Applications

Opinions have always been important to people as opinions influence people's behaviours. Commercial organizations look for the user's opinion about their product or service; similarly, consumers also want to know about the services and the support given by some company or organization, and for that users look for the opinions of others. To acquire the opinions of others, commercial units conduct surveys, polls or use forum websites, whereas consumers obtain opinions from their friends, families or from websites. Since the growth of social media, individuals and organizations have been using them as a platform to form their opinion by analyzing the sentiments of other users. The Web has played a great role in opinion mining and making opinions, as it has provided a common platform for organizations and consumers to express their views and get the opinions of others. Now, information can be filtered out by companies from these reviews and used to decipher the emotions of consumers. With the current trends, this is the best method of collecting opinions owing to a large user base giving opinions. Opinions can be either positive or negative; thus, a method of analyzing these opinions is to be formulated. Here are some simple examples where opinion mining is used:

- Business management and organizations are interested in opinions

  - Product and service benchmarking
  - Market intelligence
  - Survey on a topic

- An individual person is interested in a second person's opinion when

  - Purchasing a product
  - Using a service
  - Other decision-making tasks

- Advertisement placing service places advertisements in user-generated content

  - Placing an advertisement from a competitor if a use criticizes a product

4   Problem Description

There is an abundance of opinions, which makes it hard to take a final decision based on them and, fake and repeated opinions are common. These opinions are not structured, some are sarcastic in nature, some give a mixture of objective and subjective, whereas some give a neutral opinion. But in the case of opinions given on some e-commerce websites or tech blogs, we usually obtain structured opinions with positive and negative opinions separated. A method is needed that can determine the fake, repeated or absurd opinions and gives a well-defined idea about the product or service in terms of its polarity.

4.1   Mathematical Description

In this section, we describe the problem of opinion mining from travel websites mathematically.

Example "(1) I bought a Zopo Notebook laptop. (2) I am in love with it. (3) The build quality is amazing. (4) The battery life is also very good. (5) However, my friends think it is an outdated model."

From this review, we notice a few important points:

The review has several opinions, both positive and negative, about the Zopo Notebook laptop. Sentence (2) expresses a positive opinion about the Zopo notebook laptop. Sentence (3) expresses a positive opinion about its build quality. Sentence (4) expresses a positive opinion about its battery life. Sentence (5) expresses a negative opinion about the model of the laptop. From these opinions, we can make the following important observations:

Observation: (a) An opinion consists of two key components: a target g and a sentiment s on the target, i.e. (g, s), where g can be any entity or aspect of the entity about which an opinion has been expressed, and s gives the polarity or rating of the sentiment. For example, the target of the opinion in sentence (2) is the Zopo Notebook laptop, and the target of the opinion in sentence (3) is the sensor quality of the Zopo Notebook laptop. The target is also called a topic in the literature. (b) This review has opinions from two people, who are called opinion sources or opinion holders. The holder of the opinions in sentences (2), (3) and (4) is the author of the review ("Mark Zuckerberg"), but for sentence (5), it is the friend of the author.

Definition 4.1 Opinion: An opinion is a quadruple, (g, s, h, t), with g as the target, s as the sentiment, h as the opinion holder and t as time of the expression.

The target can be made systematically in a structured manner, which eases mining and then the uses of the mined results. For example, the *build quality of laptop* can be represented after being paired into an entity and an attribute of that entity, *(Zopo-Laptop, build-quality)*.

Definition 4.2 Entity: An entity e is a product, service, topic, issue, person, organization, or event. It is described as a pair, e: (T, W), where T is a set of parts, sub-parts and so on, and W is a set of attributes of the entity.

For example, a model of laptop is an entity, e.g. the Zopo Notebook laptop. It has a set of attributes, e.g. build quality, size, and weight and a set of parts, e.g. screen, RAM and battery. The battery has a few attributes of its own such as battery life and its weight.

Opinion as a quadruple is not complete as it does not consider all aspects of an opinion; thus, a stronger definition of opinion is needed.

Definition 4.3 Opinion: An opinion is a quintuple, $(e_j, a_{jk}, s_{ijkl}, h_i, t_l)$ where e_i is the name of an entity, a_{ij} is an aspect of the entity, s_{ijkl} is the sentiment on aspect a_{ij} of the entity e_i, h_k is the opinion holder, and t_l is the time when the opinion is expressed by hk.

The problem lies in the fact that there are mixed opinions in a document, and with the varying nature of natural languages, people express an opinion in different ways, that too without using any sentiment lexicons. The motive is to overcome this issue related to sentiment mining [16].

5 Social Network

A social network is an online service or site that concentrates on aiding the building of social networks or social connections between people who, for example, have common interests, activities, backgrounds, or real-life connections. It took a very little time for certain sections of various social networking sites to start targeting niche members. Every day we are greeted by a host of new social networks vying to catch attention. We cannot avoid social media; it is everywhere and it is making a major impact on the travel industry. Many websites have been launched in the past several years, and all of them are trying to attract travellers and take a portion of the online travel market. For many travellers, the media have evolved into an invaluable tool for every part of planning and going on a trip: finding inspiration on Pinterest, hunting down deals on Groupon, getting restaurant tips on Zomato.com and sharing photos on Instagram. For others, it is a bewildering blur of Tweets, posts and check-ins. Tips from friends are still among the best types of travel advice, and it is simple enough to post a question on Facebook or Twitter asking for suggestions on, say, a roadside Dhaba in Murthal or a waterpark in a nearby city. Not just questions, people share their reviews and express their feelings openly, giving more choices for places to visit. Everyone these days uses smartphones, which helps to connect to people with many social networking applications such as Snapchat, Instagram, Facebook, Hike, Twitter and many more. Snapchat has made a place on the social networking market using its 10-s advertisements, especially with its emphasis on visuals and immediacy, which leads its users to share stories about their travels on the platform. Another application is Instagram. Instagram is an application used for sharing photos and videos with friends. Travel brands have been using Instagram for a few years now, and it is no wonder if brands within the travel industry continue to invest in it, as Instagram itself has introduced paid advertisement facilities, adding to the benefits of advertisers. There are many social networking websites, and the travel industry is using it to its fullest extent. In this industry, Facebook is among the top social networking websites used for advertising and promotional purposes. Facebook took over the internet community very speedily. Facebook has an influence on almost every business sector, and its absence is unimaginable. The travel industry is no exception. Here, travellers' post their observations, experiences, thoughts, and interact with fellow travellers, criticizing or praising various aspects of travelling. With the development of smartphone applications in social media, each social networking site is launching their individual application that has certain features, which include embedded advertisements. Travel agencies have started to use it to their benefit; apart from this, travel websites have also launched their individual applications on platforms such as Android, iOS or Windows, and a major chunk of redirects are from these applications. Social networking giant Facebook has started a whole new trend by introducing the concept of Facebook Ads, through which the company provides vendors with an advertising facility. Travel websites use these ad panels to post a link to their websites and show posters with attractive offers and discounts. In fact, travel is Facebook's most popular subject, as in 2011, Facebook created a team of employees whose jobs are to liaise with travel brands. McCabe oversees this group, which he says was created: *to better understand brands' needs, learn to speak their language, and most importantly, build the right products and services for the industry. We verticalized for several industries, not just travel, but travel is one of the fastest growing verticals.* Travel websites have dedicated pages that hire people who like to travel, collect experiences and post them to attract customers. In addition to Facebook, Instagram also has a huge community of travellers sharing photographs & videos.

6 Travel Websites

Traveling means to make a journey from one place to other places and a travel website is a website dedicated to traveling, such as TripAdvisor.in, as shown in Fig. 10.

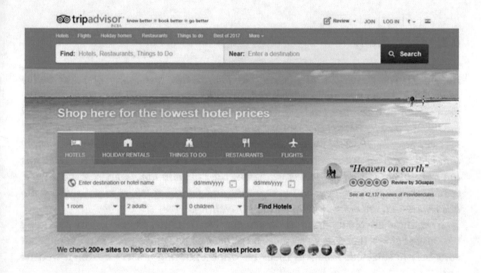

Fig. 10 Example

It shows information/suggestions/opinions from travellers and allows visitors to take more informed decisions. While collecting the information, blogs or social networking websites are referred to for a better and wider range of suggestions, and special websites dedicated to providing all that we need to know about travelling and the places to visit anywhere in the world are available. As travel and tourism are still considered to be a sunrise industry, nowadays social networking sites have had a considerable impact on them. Travel websites look at how travellers are using the major social networks plus the apps, sites and services that seem to launch every day [11]. A website with a large pool of travel enthusiasts is TripAdvisor.com. TripAdvisor, a travel community with 10 million plus reviews, typically posted by different users, of various places, who are eager to share their insights.

6.1 Case Study

Budget Travel recently sent a writer to Istanbul to test many of these online tools. With a smartphone in each hand, Arianne Cohen tested the limits of online networking in a foreign land. No guidebooks and no language skills. Only social media and mobile apps. (Her story: "The Connected Traveller.") "It was fun, but it was not easy". One tool Cohen said worked well was InterNations.org. She wrote that it is a global expat community with 230 local branches. In hard times, this is the right place to find the locals who know the common language, sharing advice, reading hotel reviews, posting photographs and finding a travel ally. Social media has remoulded the conventional ways pf travelling.

Now, there are plenty of travel sites on the Web that assist travellers by providing various services; these services must be refined as per one's need and the one that suits best is selected. Points to focus on while choosing a service from a website are:

- Packages: A package comprises a pre-planned trip with attractive offers, which can be purchased from the site.
- Travel ease: This refers to the ease of using the tools that the travel website provides, such as choosing a flight or train, allowing individuals to select the best travel dates, type of seat, the number of passengers and to explore other options.
- Rental facilities: Some people like to travel in their own way and have fun on their own. Therefore, a rental car may be needed, and this consists of the search and sorting tools needed to get the desired deal on a rental car.

- Trip selection: This is selecting the kind of deal a consumer is looking for and the kinds of deals available on the website. Travel sites also have deals on rented cars, hotels and popular local activities.

7 Experiment

In this section, we discuss experimental work.

7.1 Objective

The objective of our experiment is to perform opinion mining (sentimental analysis) on reviews extracted from a travel website, TripAdvisor.com. We trawled reviews from TripAdvisor, as it is one of the largest travel websites in the world, with more than 60 million members and over 170 million reviews for hotels, restaurants, attractions and other travel-related businesses [26]. We implemented a web crawler in Python to extract the data, i.e. reviews from TripAdvisor, which we achieved by web crawling. We applied various machine learning algorithms, such as OneR and Random Tree, on the extracted reviews from TripAdvisor.com. We evaluated the performance of [17–23] algorithms, and used various statistical measures to determine the best algorithm in mining opinion from these reviews.

7.2 Web Crawling

A web crawler (also known as a web spider or web robot) is a program or automated script that browses the World Wide Web in a systematic, automated manner. This process is called web crawling [12]. Visual representation of the process of web crawling is explained in Fig. 11.

Figure 11 shows that a web crawler starts with a list of URLs to visit, and as the crawler parses through these URLs, it identifies all the hyperlinks in the page and adds them to the list of URLs to visit. URLs from there are recursively visited according to a set of defined policies.

7.3 Collection of Data

To collect the data (reviews) we implemented a web crawler. To do this, we used an open source web crawling framework Scrapy. In contrast to web crawling, web scraping focuses more on changes in unstructured data on the Internet, and

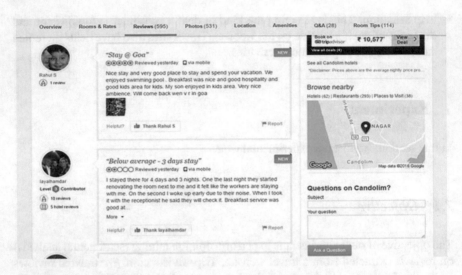

Fig. 11 Web crawler working diagram

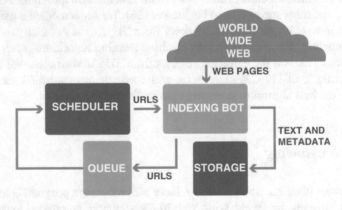

Fig. 12 Scraping process with Scrapy framework

in structured data that can be saved and examined in a central local database or spreadsheet. Web scraping is connected to web automation, which simulates human browsing using computer software. Uses of web scraping include online price comparison, contact scraping, weather data monitoring, website change detection, research, web mashup and web data integration. To extract data from TripAdvisor.com, we implemented a spider program in Python to work with Scrapy.

Figure 12 shows the scraping process using the Scrapy framework. Spiders are classes that describe the ways a website or a group of websites are to be scraped, incorporating different methods of executing the crawl (i.e. follow links) and how to obtain structured data from their pages (i.e. scraping items). In simple words, spiders are the part of code in which we define the desired functions for crawling and parsing pages of the target website (or, in some cases, a group of websites) [13]. The scraping cycle of a spider is shown in Fig. 12:

- It starts by generating the initial request to crawl the first URL. Also, a callback function is defined that is to be called with a response downloaded from those requests. The initial requests to perform are obtained by calling the start_requests() method that (by default) produces a request for the URLs defined in the start_urls() and the parse method as the *callback* function for the Requests.
- The response is parsed in the callback function and is then returned either directly with the obtained data, request objects or an iterable of these objects. These requests perform a necessary *callback* procedure (maybe the same) and will then be processed and downloaded by Scrapy, and their response managed by the specified *callback*.
- In *callback* functions, the page contents are parsed using the selectors and items are generated from the parsed data.
- Finally, the items returned from the spider are usually stored in a database through an item pipeline or written to a file using feed exports.

7.4 Identifying the Target

Before we start scraping the data, we first have to locate and identify the target data on the web page and find its identifier, i.e. a CSS selector that was defined by the website developers to distinguish that part from other content on the web page. Therefore, to find CSS selectors, we used SelectorGadget [14], an extension for Chrome Browser. The CSS selector found for the reviews field is ".partial_entry" as seen in Fig. 13.

Fig. 13 Using the selector tool and finding CSS selector for the reviews field

7.5   What Should be Extracted?

While extracting the data, we identified the rating of the reviews so as to classify them as positive or negative. While doing so we found text representing the ratings and we applied regular expressions to extract them. Our implementation of a spider automated the extraction of the links of hotels from the first page going through every single link and extracting reviews of those hotels up till the last page.

7.6   Running the Implemented Spider

We ran our spider and collected around 2000 positive reviews and 2000 negative reviews into a CSV file.

7.7   Datasets

This section includes details of the dataset used in the experiment. Also, we compared the following machine learning algorithms J48 [21], logistic model tree (LMT) [22], random forest [24], random tree [24], decision stump [20], naive Bayes [23] and OneR [1].

- J48: A predictive machine-learning system that decides the objective value of new instances based on different attribute values of the accessible data is the J48 decision tree. The different attributes are informed by the internal nodes of a decision tree; the branches between the nodes tell us the possible values that these attributes can have, whereas the terminal nodes tell us the final value of the dependent variable [21].

  Disadvantages: The run-time complexity of the algorithm matches the tree depth, and cannot be larger than the number of attributes. The size of the tree is linked to the tree depth. The size of C4.5 trees enlarges linearly with the number of attributes. The drawback of the C4.5 rules is that they are slow for large and noisy datasets, and the space complexity is enormous as we have to store the values repeatedly in arrays [5].
- LMT: A sorting model with respective supervised training algorithm that combines logistic prediction and decision tree learning is a logistic model tree (LMT). A decision tree is used in the logistic model tree that provides a section-wise linear regression model [22].
- Random forest: Random forests are an aggregate of tree predictors such that each tree depends on the contents of a random vector examined autonomously and with the same frequency for all trees in the forest. This approach is also used for equating various complex decision trees by training them on several parts of the similar training set, with the purpose of overcoming the variance [24].

- Random tree: A random tree is a combination of tree predictors that is called a forest. It can deal with both classification and regression problems. The classification works as follows: each random tree classifier uses the input feature vector and matches it with all the trees in the forest, and produces the class label that gained the majority of votes. In the case of regression, the classifier response is the mean of the responses of all the trees in the forest. All the trees are trained with the same parameters, but on several training sets [25].
- Decision stump classification algorithm: A decision stump is a machine learning model consisting of a one-level decision tree with one internal node (the root) that connects terminal nodes (on its leaves). Decision stump works by predictions based on the value of single input features. For nominal features, a stump build with a leaf for each possible feature values or a stump with two leaves, one of which corresponds to some preferred category and the second to the rest of the categories [20].
- Naive Bayes algorithm: It is a probabilistic and supervised classifier devised by Thomas Bayes [23]. According to this model, if there are two events, say $E1$ and $E2$, then the conditional probability of the existence of event $E1$ when $E2$ has already occurred is given by the following mathematical formula: $P(\frac{E1}{E2}) = (\frac{[P(\frac{E2}{E1})*P(E1)]}{E2}))$.
- OneR algorithm: OneR stands for "One Rule", a simple and accurate algorithm that generates one rule for each predictor in the data and selects the best rule. It constructs a frequency table to create a rule for each predictor [1].

7.8 Hotels Review Dataset

We trawled the reviews of hotels from "www.tripadvisor.com" of three countries (India, Italy, China). We randomly selected 2000 positive reviews and 2000 negative reviews using Scrapy.

7.9 Attributes from the Reviews

We used these reviews to find attributes. An attribute is defined as a lexeme that determines the properties of a field, a tag in a database. We scraped lexemes, from the reviews, through our implementation, to find the polarity of a review.

Example: Through the following review extracted from TripAdvisor we highlight the attributes:

Review: "A very trendy, intimate boutique hotel. Hardware is in excellent condition. Service is excellent; very attentive. The only downside is no hotel gym; they do provide dumbbells in the room, but it is still a bit strange working out in the room. Nevertheless, it is far better than the other choices in Rome, which tend to be run-down."

Attributes from the review may be trendy, intimate, boutique, strange, better, excellent, attentive.

7.10 WEKA

Waikato Environment for Knowledge Analysis (WEKA) [19] is the software of machine learning algorithms for data mining projects. These algorithms can be applied to a dataset either directly using the GUI of WEKA or using JavaScript. WEKA contains tools for data pre-processing, regression, association rules, classification, clustering and visualization. It is also relevant for developing new machine learning schemes. We applied some of these algorithms to our dataset.

7.11 Statistical Measures

In this section, we explain the statistical measures used to evaluate the efficiency of machine learning algorithms.

We used the following statistical measures to compare the machine learning algorithms in WEKA.

7.11.1 Kappa Statistic

Cohen's kappa coefficient is a performance measure of inter-rater concordance (it gives a score of how many similarities there are in the ratings provided by judges) for qualitative items. Kappa is computed from the observed and expected incidences on the diagonal of a square confusion matrix. The observed proportionate agreement is: $Po = [\frac{(a+d)}{(a+b+c+d)}]$.

To calculate Pe (the probability of random agreement) we note that:

(Class)/(Reader A): marginal $a = [[\frac{((a+b)*(a+c))}{(a+b+c+d)}]]$

(Class)/(Reader B): marginal $b = [\frac{((c+d)*(b+d))}{(a+b+c+d)}]$

Thus, the overall probability of random agreement is $Pe = [\frac{(\text{marginal } a+\text{marginal } b)}{(a+b+c+d)}]$.

$$\text{Kappa statistic} = k = \left[\frac{(Po - Pe)}{(1 - Pe)}\right]$$

If the raters are in complete agreement, then $k <= 1$. If there is no agreement among the raters other than what would be expected by chance (as given by Pe), $k <= 0$.

Table 1 Kappa statistic values for classifiers

Classifiers	J48	LMT	R. forest	R. tree	Decision stump	Naive Bayes
Kappa statistic	0.659	0.7351	0.7651	0.4513	0.3645	0.6931

Table 2 Confusion matrix

J48		LMT		RF		RT		DS		NB		JRIP		OneR		Class
815	177	886	106	913	79	759	233	923	69	869	123	790	202	923	69	Negative
206	802	159	849	156	852	316	692	569	439	184	824	222	786	569	439	Positive

Table 1 shows the values of the kappa statistics, a performance measure of inter-rater agreement. It uses the confusion matrix, from Table 2 to find out the value for each classifier. Table 1 shows that random forest performs better in terms of kappa statistics than other algorithms.

7.11.2 Confusion Matrix

The confusion matrix is also known as the contingency table. Here, we have two categories; thus, we kept a $2 * 2$ confusion matrix; the matrix could be large if we have more than two categories. The number of correctly classified instances are the sum of diagonals in the matrix. A confusion matrix has the following outcomes:

TP = True positives: number of instances predicted positive that are actually positive.

FP = False positives: number of instances predicted positive that are actually negative.

TN = True negatives: number of instances predicted negative that are actually negative.

FN = False negatives: number of instances predicted negative that are actually positive.

In the confusion matrix (Table 2), diagonal elements (i.e. true positive and true negative) are divided by the total number of instances in the negative and positive classes respectively, i.e. $[\frac{815}{(815+177)}] = 0.822$ for the negative class and $[\frac{802}{(206+802)}] = 0.796$ for the positive class, in our case for the J48 algorithm.

The false positive (FP) rate is the proportion of the attributes that were classified as the positive class, but belong to the negative class, as shown in Table 2. In the confusion matrix table, this is the number of attributes classified as negative in a positive class divided by the number of instances in a negative class. i.e. $[\frac{(177)}{(1008)}] = 0.177$.

The true positive (TP) rate is the proportion of attributes that were classified as the positive class, as shown in Table 3, among all attributes that truly have a positive class, i.e. what proportion of the class was captured out of that class, is equivalent to recall.

Table 3 Outcomes of the confusion matrix

	TP rate	FP rate	Precision	F-measure	Recall	Class
J48	0.822	0.204	0.798	0.81	0.822	Negative
	0.796	0.178	0.819	0.807	0.796	Positive
LMT	0.893	0.158	0.848	0.87	0.893	Negative
	0.842	0.107	0.889	0.865	0.842	Positive
Random	0.92	0.155	0.854	0.886	0.92	Negative
Forest	0.845	0.08	0.915	0.879	0.845	Positive
Random	0.765	0.313	0.706	0.734	0.765	Negative
Tree	0.687	0.235	0.748	0.716	0.687	Positive
Decision	0.93	0.564	0.619	0.743	0.93	Negative
Stump	0.436	0.07	0.864	0.579	0.436	Positive
Naive	0.876	0.183	0.825	0.85	0.876	Negative
Bayes	0.817	0.124	0.87	0.843	0.817	Positive
OneR	0.93	0.564	0.619	0.743	0.93	Negative
	0.436	0.07	0.864	0.579	0.436	Positive
JRIP	0.796	0.22	0.781	0.788	0.796	Negative
	0.78	0.204	0.796	0.788	0.78	Positive

The precision is the proportion of the attributes that truly have a negative or positive class among all those that were classified as negative or positive class in Table 2. Table 3 gives values for each instance when we divide its true value by the sum of its true value and the false predicted value. Precision $= [\frac{(\text{True value})}{(\text{Truevalue}+\text{Falseprediction})}]$.

The F-measure is simply $[\frac{(2*\text{Precision}*\text{Recall})}{(\text{Precision}+\text{Recall})}]$, a combined measure for precision and recall. In the confusion matrix, the F-measure for the negative class is $[\frac{(2*0.821*0.843)}{(0.821+0.843)}] = 0.843$ shown in Table 3.

7.12 Instances of Classifiers Used

The first $\frac{2}{3}$ data are used as a training set and the rest of $\frac{1}{3}$ as a testing set. Table 4 shows that random forest is the best algorithm so far, which correctly classified 88.25% of instances and OneR gives the worst results with 68.1% correctly classified instances.

7.13 Outcomes of the Confusion Matrix

After applying one of the above testing options, we now use the classifier algorithm. Eight algorithms are implemented on the dataset, as stated in Table 5.4. This shows

Table 4 Instances of classifiers used

Algorithm	J48 (%)	LMT (%)	R. forest (%)	R. tree (%)	D. stump (%)	N. Bayes (%)	JRIP (%)	OneR (%)
Correctly classified	80.85	86.75	88.25	72.55	68.10	85.65	78.80	68.10
Incorrectly classified	19.15	13.25	11.75	27.45	31.90	15.35	21.2	31.90

the percentage of the instances that are correctly classified and those incorrectly classified. WEKA also uses the confusion matrix to provide a clear view of these classifiers. Table 5.2 gives the values of the confusion matrix for each classifier, it contains four values in each entry as: (1) true positive; (2) true negative; (3) false positive; (4) false negative. Both (1) and (2) predictions are exact as they exist, but (3) and (4) predicted the opposite of their properties. Table 5.1 shows values of the kappa statistics. It uses the confusion matrix, from Table 5.2, to find out the value for each classifier. Table 5.3 gives values of: (1) true positive rate; (2) false positive rate; (3) precisions; (4) recall; (5) F-measure. Precision is the fraction of retrieved instances that are relevant, whereas recall is the fraction of relevant instances that are retrieved. Table 5.3 shows that: The TP rate gives the best result in a negative class for decision stump and OneR, which is 0.930, and for a positive class, it gives the best result for random forest (0.845). Recall is also equivalent to TP rate; thus, it gives the same result. The FP rate gives the best result in a negative class for decision stump and OneR, which is 0.564, and for a positive class, it gives the best result for random tree (0.235).

Precision gives the best result for the random forest for both negative and positive classes, i.e. 0.854 for a negative and 0.915 for a positive class. F-measure gives the best results for both a negative and a positive class for random forest, i.e. 0.886 for a negative and 0.879 for a positive class.

7.14 WEKA Classifiers

WEKA is used to implement these algorithms and they provide several testing options such as: (1) use training set; (2) cross-validation; (3) percentage split. Option (1) uses a pre-compiled dataset as a training set, and by using this, it classifies instances of the testing set. Options (2) and (3) use the same dataset for both training and testing. In Table 5.4 the percentage split of the dataset is as follows: the first $\frac{2}{3}$ is a training set and the rest of $\frac{1}{3}$ is a testing set divided randomly.

8 Conclusion

In this chapter, the concept of opinion on sentiment analysis has been defined. Two important concepts, subjectivity and emotion, were also introduced. We sought to classify sentiments at a different level, namely: (1) document level, (2) sentence level and (3) feature/entity level. We noticed the limitations of these classification levels, for example, classification at the document level could only be done if the document has reviews for a single entity only because multiple entities cannot be clubbed together under the same category. In practice, it is the forum, discussions and blogs that need sentiment classification to determine people's opinions about different entities and topics. Sentence level classification goes deeper than the document-level sentiment classification, as it moves closer to opinion targets and sentiments on the targets. Although many efforts have been made in the research community regarding this problem, and many systems have been built, but the problem is still far from being solved. Every sub-problem remains highly challenging. The two basic problems of entity level classification are aspect extraction and aspect sentiment classification. We extracted data from TripAdvisor.com/TripAdvisor.in, as it is the most famous and trusted travel website in the world. We compared eight classification algorithms of WEKA. Random forest shows the best result that correctly classifies 88.25% instances and 11.75% incorrectly classified instances, and this is because of the structure of random forest, as it has a combination of tree predictors, where each tree predictor depends on the value of the random vector independently. With a single predictor, OneR gives the worst results, yielding only 68.1% correctly classified instances and 31.9% incorrectly classified instances.

9 Future Work

Ensemble learning is a procedure that is applied to combine and train multiple algorithms and combining their outputs to obtain a better result with increased accuracy and reliability. Ensemble methods have a multi-disciplinary approach and are being used in bioinformatics, medicine, finance, manufacturing, information security, information retrieval and image retrieval, etc. Further research is required to develop an ideal ensemble technique that can achieve high accuracy and usability. Also, a method for combining outputs of classifiers should also be developed, with a focus not only on optimal performance, but also on better decision-making capability [15]. The applications of data mining are part of human life in various fields, such as business, sports, higher education, the medical and scientific fields, politics, and in security and preservation of privacy. An example for the recording of electronic communication such as email logs and web logs have captured the human process. Some of the application areas of applying ensemble techniques for opinion mining include:

- Mining financial trade, for example, using refined data of asset prices to gain greater profits.
- Detecting natural disturbances.
- Sports trading: One of the most popular sports in India is cricket. Even though sports trading is illegal, traders have a large illegal network that operates from the shadows and they use social networking to stay updated about various aspects of sports persons.
- Manufacturing or warranty analysis: Companies examine the text from warranty claims, technician's reports, order requirements, customer relations texts, and other necessary information using text analytics to extract certain entities. They can then analyze this information, and determine the necessary actions.

References

1. Bayes, N.: C-4.5 algorithm. Naive Bayes classifiers (2015)
2. Bogdan, C.B.T.: Solutions for working with large data volumes in web applications. Bucharest (2011)
3. Frand, J.: Data Mining: What is Data Mining? UCLA Anderson School of Management, Los Angeles, CA
4. Giatsoglou, M., Vozalis, M.G., Diamantaras, K.I., Vakali, A., Sarigiannidis, G., Chatzisavvas, K.C.: Sentiment analysis leveraging emotions and word embeddings. Expert Syst. Appl. **69**, 214–224 (2017)
5. Juneja, D.: A novel approach to constructing decision tree using the quick C4.5 algorithm. Orient. J. Comput. Sci. Technol. **3**(2), 305–310 (2013)
6. Liu, B.: Web Data Mining - Exploring Hyperlinks, Contents, and Usage Data. Springer, Berlin (2006)
7. Liu, B.: Sentiment Analysis and Opinion Mining. Morgan & Claypool Publishers, San Rafael, CA (2008)
8. Meire, M., Ballings, M., Van den Poel, D.: The added value of auxiliary data in sentiment analysis of Facebook posts. Decis. Support. Syst. **89**, 98–112 (2016)
9. Mureddu, D.O.F.: Research challenge on opinion mining and sentiment analysis (2008)
10. Piryani, R., Madhavi, D., Singh, V.K.: Analytical Mapping of Opinion Mining and Sentiment Analysis Research During 2000–2015. Elsevier, Amsterdam (2016)
11. Ratan, R.: A proposed novel approach for sentiment analysis and opinion mining. Int. J. UbiComp, **5**, 1 (2014)
12. ScienceDaily: Web Crawler. ScienceDaily LLC
13. Scrapy-Developers: Spiders. Scrapy.org (2010)
14. SelectorGadget-Developers: SelectorGadget: point and click CSS selectors. SelectorGadget.com (2010)
15. Shahid, S.A.: Ensemble Learning Methods for Decision Making: Status And Future Prospects. IEEE Xplore: Digital Library (2015)
16. Similjan, T.: How Social Media is Changing the Travel Industry. Time Inc. Affluent Media Group, New York (2016)
17. Smeureanu, I.: Applying supervised opinion mining techniques on online user reviews. Inf. Econ. **16**(2), 81 (2012)
18. TripAdvisor.in: TripAdvisor - know better, book better, go better. TripAdvisor LLC. (2016)
19. University of Waikato: WEKA Documentation
20. University of Waikato: Decision stump algorithm. weka (2015)
21. University of Waikato: J48 algorithm. weka (2015)
22. University of Waikato: LMT algorithm. weka (2015)
23. University of Waikato: OneR algorithm. weka (2015)
24. University of Waikato: Random forest algorithm. weka (2015)
25. University of Waikato: Random tree algorithm. weka (2015)
26. Wikipedia: TripAdvisor: From Wikipedia, the free encyclopedia. Wikipedia LLC (2010)

Facilitating Brand Promotion Through Online Social Media: A Business Case Study

Mini Ulanat and K. Poulose Jacob

Abstract Online social networks promote faster propagation of new ideas and thoughts. The percolation of social media into our daily life has influenced the way in which consumers interact and behave across the world. Social media tools are emerging as powerful alternatives to the traditional media. With more than two thirds of the internet population connected through online networking sites such as Facebook, Twitter and MySpace, the potential offered by this medium is tremendous. This chapter is an attempt to analyse the case studies of three brand promotion activities through Facebook and to model the mechanics of successful diffusion in comparison with traditional channels.

Keywords Brand promotion • Information diffusion • Online social network • Social media

1 Introduction

The growth of the Internet and the use of online social networks have led to a new collaborative platform for communication and information exchange. This technology has revolutionised the pattern of production and consumption of information. Leading to more customer interaction, this became the mainstream media, thereby improving visibility. Anonymity is an encouraging factor for virtual interactions. Online social network sites have brought in a change in the communication, positioning themselves as unchallenged in sharing multimedia content in addition to private messages and public blogs. Social media includes all channels of communications and social media tools are increasingly being used to connect dispersed individuals. This technology has basically changed the way we interact in everyday life, our entertainment habits, our shopping decisions to the information dissipation in society, which also has a direct impact on our democratic election process.

M. Ulanat (✉) • K.P. Jacob
Cochin University of Science and Technology, Cochin 682022, India
e-mail: mini.ulanat@gmail.com; kpj@cusat.ac.in

© Springer International Publishing AG 2017
H. Banati et al. (eds.), *Hybrid Intelligence for Social Networks*,
DOI 10.1007/978-3-319-65139-2_9

This chapter is structured as follows. Starting with a little history and growth of social networks and definitions of network terms in Sect. 2, we outline the theoretical background of the mathematical concepts of social networks in Sect. 3. The data collection process and analysis are also described. Social network analysis techniques and how they are used to gain insights into the dynamics of the information flow among them are presented in Sect. 4. Conventional and digital modes of social interaction are discussed in Sect. 5 and a few case studies in Sect. 6. Highlights of the development and application of these test cases are documented and are discussed in more detail. The conclusions and reflections on data analysis for future research are presented in Sect. 7.

2 History, Growth and Application of Social Networks

Social networks can be defined [1] as networks with people as nodes and their relationships represented as edges connecting the nodes. Social network analysis (SNA) is the mapping and study of relations in the social network to understand the behavioural patterns. The appearance of online social networks has revolutionised the world wide web. Advances in web sciences have helped to provide comprehensive digital traces of social actions, interactions and transactions that made the analysis a simpler task.

The history of social networks dates back to the Stone Age where humans gathered around campfires and shared stories or painted history on cave walls. These activities were influential and the tools they used, including words, paints, signals, smoke, stone arts etc. were their social media. Human nature is always to be a part of society and be connected. The communities have always evolved around an area of common interest in the long history of mankind.

With the fast pace of changes in the computer hardware and communication technologies, connectivity and networking became easy and affordable. These virtual networks facilitate social connectivity by creating a web profile and thus managing their identities. The internet made it possible to reach around the globe with mere click of a button. The world wide web shifted the information-sharing medium to a communicative medium. The recent practice of connecting people through their web presence and encouraging group discussions helped to bring an affordable internet service and thereby online conversations to the mainstream [2]. Thus, the online social networks during the period 1978 to 1994 played a crucial role in promoting a new communication culture. The history of social networks can be linked to chat rooms. The online network in its primitive form appeared in 1995 with the site classmates.com for keeping in touch with schoolmates and the modern form took off with Friendster in 2002. The popularity of Friendster [3] was so great that its user base grew to 3 million users in the first 3 months. In 2003, MySpace [4], a Friendster clone, was launched. MySpace had a good customer base for a few years, but had to give way to Facebook, which was launched in 2004 as a closed media and became public later, in 2006 [5].

2.1 Social Network Analysis

Social network research, which started with a sociological experiment of sending mails across the country through first-level contacts only, evolved into the famous theory of Six Degrees of Separation [6]. The analysis of social interactions as a network has its origins in the early twentieth century, with studies originally concentrating on limited centres from a sociological perspective. With the advent of speedy internet and the large amount of data available, it is now possible to use faster methods of data analysis.

Popular applications such as YouTube, Flickr, MySpace, Facebook etc. have millions of active users and provide us with many online social networking possibilities. Networks such as Facebook [7], Twitter [8] and LinkedIn [9] have quickly become de facto communication platforms, with large amounts of social networking data, whose analysis has become a challenging task. New mathematical models and analysing tools are increasing being applied to understand how online social networks are formed and how the network topologies are efficiently shared and information distributed. These models provide surprising insights into how the social networks are formed and efficiently utilised in everyday life. Although investigations into social networks started many years back, the scope of the analysis was limited owing to the small data samples collected using questionnaires and interviews. Currently, the colossal data generated through online social networking sites, consisting of millions of nodes and connections, can be extracted and are available for analysis. These data are extremely difficult to analyse through traditional methods because of the scale, complexity and dynamics of the networks involved. With the big data analytics improving, the available data about human communication, common activities and collaboration has opened up new vista that provide new insights. SNA relates to mapping, understanding and analysing interactions across a set of people.

Social network analysis (SNA) [10] implements an experimental evaluation mode of message exchanged among the members of the SN, identifying the community structure. SNA is becoming the standard approach for analysing this network evolution process. It helps in understanding the patterns of social relationships among the entities in an organisation. Monitoring and analysing data generated by social channels have emerged as a novel branch of social intelligence. The user-generated contents in the SN are in the form of *likes*, *comments*, *posts* etc. In the SN interactions, both strong and weak links are important. The information flow through socially weak ties can bring in more benefits extending outside of normal relationships.

2.2 Business Applications of SNA

Social network analysis applies to a large array of business problems, involving inter-personal relations and inter-organisational relations. It helps in identifying key influencers, understanding their behaviour and studying how these networks change over a period of time. Some of the important business applications of SNA include:

1. Knowledge Management: Improved information flow and knowledge-sharing, building communities of practice. Improving inter-organisational relationships
2. Customer Behavioural analysis
3. Social media
4. Health and social care
5. Recommender system: collaborative filtering and personalisation
6. Opinion modelling
7. Crime detection and investigation
8. Economic predictions
9. Political Influence Modelling

3 Mathematical Concepts of Social Networks and Social Network Analysis

The quantifiers generally used in the analysis of social networks are centrality, closeness and betweenness. The importance or influence of nodes and edges are gauged by these measures. Even though many studies have been carried out using these measures, they are mainly limited to shorter datasets arising out of small interactions. The above studies have helped in the tracking of how people share information and what kind of information is spread, and the reach of the spread. Online social networks encouraged predefined interactions through this network. The two most important factors considered in promoting the brand are influence and relevance.

A social network is the structure of the community including all interactions between individuals of the community. These interactions are referred to relationships or ties. The social networks do not look into individual characteristics. The developments in computing methodologies and mathematical techniques help to visualise and measure network parameters. The quantitative approach of measuring and mapping network parameters and equating the social relations into numeric data helps with the understanding of the dynamics of human relationships. This has been studied for a long time by social scientists. The social relations are mapped into binary categories, thus a qualitative approach is complimenting quantitative work. The introductory level definitions and measures used in network analysis are explained in brief in the following section [11]. A social network consists of a number of units (nodes) that are connected to each other by a defined relationship, for example, in a citation network, X cites Y; in a professional or friendship network, A sends 6 email messages a week to B, or similar. There are a few general terms: the units may be persons, organisations, cities, journal articles, or other types of entities; the relationships may be uni-directional or bi-directional; and the linking relationships may represent categorical relationships or intensity relationships.

X and Y are friends is a bi-directional relationship; X and Y are close friends is a bi-directional relationship, recording the intensity or frequency of communication.

The network is called a graph in the mathematical literature. A graph is a collection of vertices joined by edges. Vertices and edges are called nodes and links in computer terminology, whereas in sociology, they are called actors and ties. A network is represented by an adjacency matrix. In a social network, connections may have weights representing frequency of contact between actors. Elements of the adjacency matrix can be given the value equal to the weights of corresponding connections. A directed graph or digraph is an edge that has direction pointing from one vertex to another. A path is a sequence of vertices such that every consecutive pair of vertices is connected by an edge in the network. A geodesic path is the shortest path between the two vertices so that no shorter paths exist. The diameter of a graph is the length of the longest geodesic path between any pair of vertices in the network, for which the path actually exists. An Adlerian path is traverses each edge in a network exactly once. A Hamiltonian path visits each vertex exactly once. A network can have one more or no Adlerian and Hamiltonian paths.

3.1 Metrics of SNA

Social network analysis focuses on patterns of relationships independently of node attributes. The quantitative metrics allow the analyst to systematically analyse, compare and track changes in the network over a period of time. The following are the conceptual definitions.

The actor/node/point/agent are social entities such as persons, organisations, cities etc. The tie/link/edge/line/arc represent relationships among actors. A dyad consists of a pair of actors and the (possible) tie(s) between them and a triad is a subset of three actors and the (possible) tie(s) among them. A subgroup is the subset of actors and all ties among them, the collection of all actors on which ties are to be measured is called a group and the collection of ties of a specific kind among members of a group is called relation. A social network is a finite set or sets of actors and the relation or relations defined on them.

3.2 Measures of SNA

The three most popular individual centrality measures of a complex network are degree centrality, betweenness centrality and closeness centrality. Degree centrality of a reference node is the number of nodes directly connected to it. In a directed network, the number of incoming links to a node are its indegree and the number of outgoing nodes from it are its outdegree. Betweenness centrality quantifies the frequency that a node behaves as a link among the shortest paths between any two nodes in the network. Closeness centrality represents the length of the average shortest path between a vertex and all vertices in the graph.

A social network is a graph of relationships and interactions within a group of individuals. This network acts as a medium for spreading information, ideas and influence among its members. Social networks build relationships and social media allow transmission and sharing of information to a broad audience built through these networks. They also open a way to the mapping and gauging of customer sentiments through active engagements. Anyone can track the content shared and measure the brand value. These social structures are modelled using the traditional graph theory of $G = [V, E]$, where persons are vertices and their social relationships are the edges.

4 Network Effects: Information Diffusion Through Social Media

An Online social network (OSN) is a highly cross-disciplinary study collaborating between Computer Science and Informatics, Sociology, Cognitive Science and Psychology, Geographic and Environmental Science, Biology, and Health and Social Sciences fundamentally powered by mathematics. Social networking sites transformed themselves into social media by engaging and creating relationships, communicating with readers, building and connecting with the online audience, abolishing boundaries with social media.

Social media radically changed information production, transfer and consumption. Promoting a brand through the successful utilisation of a social media platform is a form of leveraging the "social" through the "media" to "market" business elements without explicitly mentioning it.

The promotion of free participation in social media helped the growth of social media marketing. Social media are the new words of mouth, opening lines of communication with consumers, building relationships with them and tracking consumer behaviour. The advantage of two-way communication helped in a long-term bonding utilised by marketing channels. This transformation helped in understanding customer expectation and delivering the information in a better way, while networking simultaneously. According to social media marketing [12], the various activities integrating technology, social interactions and content creation are defined by the umbrella term social media. Having had a profound impact on our daily lives, social media became an important conduit for brands to build loyalty and sway purchase decisions. Brands realised the importance of an online presence and the direct interaction with clients, thereby gaining a market share. The methodology consists of reaching the community and entering into the conversation. Social media democratised the tools of self-expression. Organisations traditionally used paid media to promote the message. Social media empowered consumers with a wealth of information at their fingertips, leveraging their influential social contacts to make buying decisions. The popularity of online social networks such as Twitter and Facebook helped to assess the visibility of the propagation of information patterns.

The metrics helped to evaluate social media exposure, influence and engagement, thereby helping to understand and track how people share information and the type of spread of information. The link between campaigning and brand awareness can be evaluated through online engagement and real-time measurement.

4.1 Facebook as Social Media

Facebook is a relation-based online social networking site that has 800 million active users [13]. It allows the users to create a personal profile and connect with their friends. As per the statistics in 2010 [14], an average Facebook user has 130 friends. The profile and interactions of a user are archived and visible to all the friends, who can interact with the user through a comment denoted as a *like*. Facebook focuses on building and reflecting the social networking and relationships among the community sharing similar interests. Facebook has become an integral part of the human digital life style. The changes evolved as emails replaced the traditional snail mails, giving way to wall posts. With the development in communication technologies, there has been an increase in online human activities, lowering the cost. It has opened up a new vista of personal profiling for self-expression, describing who you are, your interests, keeping in touch with friends, sharing photos, videos etc. Social network sites such as Facebook, Twitter etc. are highly valued for their collected detailed user profile and network data and their retention capability. This is extremely helpful in the case of targeted advertising. Realising the potential and seeing the success, small businesses also joined by building relationships with new demographics.

4.2 Brand Awareness

Brand awareness quantifies the familiarity of a particular brand among target consumers. Advertising has a significant effect on promoting brand awareness. Social media comprise a dynamic network and a study of the diffusion of information helps to strategise advertisements. The correlation between the number of people engaging in spreading information and the probability of adopting information as an important feature of information diffusion has already been examined. The studies of the diffusion of information in various fields led to the theories built on direct observation in small networks and survey response from large samples. This idea is important in the marketing mix, to improve the relations with customers. People are connected through predefined online interactions forming a network. These networks share messages with the connected contact and propagate information through links already formed and creating cascades. By encouraging the customers to share product information with their friends via social media, viral marketing is promoted.

Evaluation of brand awareness and the popularity of a brand of a product is difficult using the traditional marketing methods. On the other hand, the online social media such as Facebook provide an opportunity to study social interactions starting from the spread of information to influence patterns and thereby the decision-making process. This decision and experience, whether good or bad, influence many others who are also in the process. The L.E.K. media consumer survey reported by Smith and Rourke [15] reveals that tech-savvy consumers are found to be 16 times more influential in the purchase decisions of family and friends. From the organisation perspective, this highly interactive nature of social media provides valuable data regarding customer preferences and behaviour from the direct and immediate feedback. This direct engagement increases brand loyalty and credibility. The case studied and examined here details the characteristics of this social interaction.

5 Promoting Brand: Conventional and Digital Modes

Business engages customers and prospects with their brand so that the engaged customer tends to purchase the brand and supports the brand by promoting it among their contacts. The business needs to engage new customers, while retaining their loyal customers. This encourages the existing customer to continue to be loyal and to purchase and support the brand by promoting it among their contacts. In social media, this is related to the community, comprising friends, acquaintances or just simple participants. The business firm always tries to keep the customers engaged in some way with the brand. This promotes the brand's popularity or the marketer's acquaintances to the world. Social media constitute a platform for content marketing and other aligned activities such as customer support, reaching out to a target audience and building the brand. The brand's popularity is reflected in the content generated through the social media port. The brand engagement helps marketers to measure the reach and consumer influence to favour their brands or to try their products [16]. In the current scenario, the vocal feedback of the consumer is treated as the best method of promotion, explaining whether a consumer is satisfied with the brand. The brand is then promoted to the friends and others in the consumer's group [5]. This is the traditional inter-personal communication. Social media have provided a new measurable format of shared engagement, unlike the conventional method.

Brand promotion is an ongoing activity. Reaching out to more people, maintaining loyalty to the brand and promoting the brand should be done regularly. Channels are important when reaching people, as people use numerous communication channels. Various media are available, but it is difficult and expensive for companies to use them all. Of all the available channels, social media brand promotion has become the de facto standard among the digital media as masses are available on these platforms and you can reach them and engage effectively. Social media such as FB have become a key item in a marketer's tool box in this digital age. This is a new and creative way of branding and leveraging the Facebook platform. Small and medium business enterprises are starting to use social media, understanding its power to reach existing and potential customers. They use social media to post a variety of texts, photos and multimedia content aimed at people from different walks of life and to drive traffic into their site.

Diffusion of information occurs in a social network owing to the spreading of items through word of mouth (WOM) and other factors. There have been many studies on the diffusion process at both macro and micro levels. The macro model focuses on the number of adopters in the diffusion process, whereas micro models determine the individual adopting pattern. In the conventional real-world scenario, there is no 100% awareness of products. Marketing firms have been trying many methods of brand promotion.

Social media are networks where the two most important factors considered in promoting the brand are the influence and relevance of friends or acquaintances or just idle participants. This demonstrates the brand's popularity or indicates the marketer's friends to the rest of the world. The fact that social media are aimed at socialisation, which is people-centric rather than brand-centric, the marketers realised that marketing strategies need to be redefined in terms of brand engagement. This should lead to consumer interaction. Unlike in a conventional WOM promotion, with social media, a new dimension in the brand promotion has been created and the shared engagement recreated in a new quantifiable format.

The Bass diffusion model [17], which is the standard diffusion model, captures the adoption of new products in a social network. This model explains the process by which new products are adopted through interaction between users and potential users. The Bass model assumes that the traditional S shape of the adoption curve is derived from the mixture of internal and external influences. Internal influences include the communications within the social system, including WOM.

An external influence comes from agents outside the social system. A combination of these two effects describes the adoption path for a new product. The model displays a cumulative S curve of adopters when the number of users of a new product is plotted against time. Initially, the adoption starts slowly, then accelerates and spreads steadily. Once the population becomes saturated, it slows down. In this paper, we are experimenting with three datasets and establishing a general probabilistic framework used to derive a macro level diffusion model, which confirms the well-known Bass model (BM). Using the case study research, the effectiveness of SN over traditional media is analysed to understand whether it conforms with the BM and also to identify implementation challenges when traditional media are replaced with social media.

6 Case Studies

Some aspects of the following case studies have already been presented at a Conference on IT in Business Industry and Government, in 2014 [18]. The rate of growth of the influence of social media is exponential, perhaps assuming the role of the most successful medium of communication among mass media. It has become the largest market place where you can reach out instantly using the right mix. The correct social media strategy can lure the customers to the product. The social media marketing analytics are able to throw light on customer attributes such as their purchasing power, buying habits, latest trend etc. This information helps to further develop better customer-friendly products. Understanding this concept, and realising that the return on investment (ROI) of this medium is better than that of other traditional media, many firms have added social media to their marketing mainstream. This realisation has given online marketing a new dimension for companies to include a social networking site as most important for their brand promotions. This mode is influential in promoting brands and products at a fast rate. In this paper, we consider Facebook as part of the social media.

There has been a growth in brand engagement with the availability of many analytical tools, tools built in to the social media and third-party tools. The brand promotion has greater implications than just recommending to others. As a first step, we define the meaning and working of customer engagement. The brand engagement is measured within a social media context. The challenge of this measurement is ROI in social media space in terms of consumer and brand engagement. With all these constraints in mind, we address the issues using the available tools.

In this paper, we design a model marketing campaign to create awareness and adoption of a new product among consumers linked by social media. We consider using Facebook. Consumers learn about the product either by receiving a direct advertisement or through WOM from their social contacts. When the network is modelled as a random graph, product awareness grows over time. It is also considered that in this process, some consumers may fail to learn about the product in the long run because they have no aware contacts to spread the word. We use social analytics tools to characterise the optimal marketing mix when the monopolist can invest in both direct brand promotion and in "viral" content designed to stimulate word-of-mouth.

Brands utilise the new promotion through Facebook pages with the following:

1. Improving social engagement: To significantly increase the success rates, promotions are launched directly onto the brand page. People are more likely to participate and add more "likes" this way.
2. Extend engagement and reach: To broaden the audience reach and overall post engagement, page posts about promotions are significantly positioned in the news feeds of those who like the page frequently.
3. Posting in the timeline: A faster and effective way of implementing promotions of a brand page for a better reach is via the timeline, as they have the ability to be spontaneous. This includes a designated weekly giveaway or any seasonal-type promotions that brands can implement as they see fit.
4. Increasing the number of fans: Depending on the promotion value, brands allow fans to have some advantage of giving ability to participate in some promos, which increases the fan base and prompts non-fans to enter to win and become a fan.

6.1   Measuring the Impact in Facebook Promotion

Facebook has analytical tools such as page insights [8] that provide details regarding a page's performance, demographic data about the audience, and that see how people are reaching, finding and reacting to the posts. Page insights regularly provide the trends, which helps to develop better-performing content. It helps us to understand who responds to the messages. This also gives the parameters gender, age and location of the people who are the most engaged with your business, which helps in continuing to engage them through targeted advertisements and promoted posts. Facebook automatically optimises a campaign so that more of your budget goes to the brand promotion that is performing the best.

6.2   Diffusion Model

The publication of the first mathematical model of new product diffusion by Bass resulted in the entry of diffusion models into the marketing discipline. Bass realised that it is possible to use diffusion theory to mimic the S-shaped growth pattern. The Bass model is one of the most popular diffusion models to describe the process of how new products are adopted, extending to a social network. In this paper, we experiment with three datasets and establish a general probabilistic framework, which is used to derive a macro level diffusion model that confirms the well-known Bass model.

There are two types of diffusion effects: innovation and imitation. Innovation is the awareness of a brand by advertising and promotions; imitation is the awareness of a brand caused by people-to-people communication. The Bass model combines innovative and imitative behaviour into one model. The model presents a rationale of how current adopters and potential adopters of a new product interact.

Where:

$f(t)$—The rate of change of the installed base fraction
$F(t)$—The installed base fraction
p—Influence parameter for innovation
q—Influence parameter for imitation

This expression can be rewritten for additional intuitive understanding using the equivalent representation. There are two special cases of the Bass diffusion model. The first special case occurs when $q = 0$, when the model reduces to the exponential distribution. The second special case reduces to the logistic distribution, when $p = 0$. The Bass model is a special case of Gamma/shifted Gompertz distribution (G/SG).

The chapter describes a case study in detail and two cases in brief relating to social media promotion designed for a marketing campaign to create awareness and adoption of a new product among consumers linked by social media. We are using

Facebook, where consumers are informed about the product either by receiving a direct advertisement or via their social contacts. The network is modelled as a random graph and product awareness is plotted against time. Social analytics tools are used to characterise the optimal marketing mix. Facebook insight tools [9] give the metrics that help to study the patterns.

For the purposes of the study, descriptive research is undertaken. The data used in this study are shared by an authorised agency involved in social media promotion. The researcher was made familiar with the domain and the problem to be studied. There are three case studies considered in this paper. The different metrics examine the increasing significance of social media as an effective tool in achieving the targeted communication objectives of organisations in the emerging economies and the effectiveness of this kind of promotion was studied earlier. The data are taken for periods of different durations with respect to different case studies.

A firm introduces a new brand into a population of consumers connected by a random network. The consumer is made aware of this brand through direct advertisement or through a social contact. Brand promotion is a continuous, ongoing process. Reaching out to society, standing by the brand and promoting the brand are done regularly. The brand awareness and product adoption move together over time. The model is dynamic.

6.3   Case 1: Brand Promotion of XYZ Hypermarket

As an example, analysis of data from Facebook pertaining to the brand awareness promotion of a department store in the local market of a tier 2 city is considered. The DS is part of a global chain of stores. The impact of information diffusion using SNA with Facebook as a medium is explored in this example. To protect the company's privacy, it is referred to as the XYZ Hypermarket. It has about 100 permanent branches around the world. The main aims of such a detailed enquiry are:

(a) To obtain an insight into the idea of SNA
(b) To comprehend the consequences of demographics in brand promotion (BP)
(c) To measure the efficiency of social media (SM) activity in developing markets

6.3.1   Research Methodology

All the data collected and analysed are sourced from a reputable agency contacted by the company itself. This gives the researchers an accurate picture and helps to understand the topic better. The increasing prominence of SM as an engine to drive the branding and awareness goals of the company in developing markets effectively in a local area is investigated in detail. The data are taken for a period of 1 year from April 2011 to March 2012, but only those from the period June 2011 to March 2012 are analysed.

6.3.2 Metrics Measured

Facebook records the followers of the brand page while also inspecting outreach to numbers of friends from those who become fans during a specific time frame or during a promotional activity, and those who interacted with the post. This gives a sense of possible monthly Facebook influence and values to quantify it.

Impact is generally considered to be a percentage of positives, negatives and neutral sentiments when applied in relation to engagement metrics and metrics for reach when applicable. Influence is a subjective metric that depends on the organisational viewpoint. Theoretically, with an increase in engagements, users are more likely to respond with action. This achieves the aim of improving brand visibility in a much easier way than before when only non-digital methods were available.

Note that in Table 1, *Likes* for the period from April to May are not shown as they are negligible.

In Facebook tracking, the number of fans for the page, the number of friends that each user has, those became fans during the month, commented on the post during the month, liked the post during the month, are all recorded. This helps to understand the reach, engagement and influence through this channel, which calculates the impact that can then be applied as a model to other social networks.

6.3.3 Brand Page Analysis: Last 12 Months

Daily Users

In the counting time period of 1 year, a visible fan count hike is only noticed in June. The fan count on the page at that time was around 2328 and this also instantly aggravated the content that was posted to the page and other activity on the page. The growth has been linear, as shown in Fig. 1.

Table 1 Likes for the period from June 2011 and March 2012

Month	New likes	Total likes
June	2328	2328
July	6682	9010
August	4083	13,093
September	7266	20,359
October	9359	29,718
November	6623	36,341
December	8001	44,342
January	3508	47,850
February	9238	57,088
March	9065	66,158

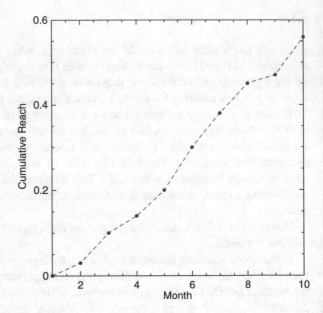

Fig. 1 Increase in the number of adopters over a period of 1 year

Likes

The "Like" option on the page is the acknowledgement that any brand page is seen. Before the timeline was introduced, the "Like" option played a very important role in fan acquisition and for engaging content. The page visit is steadily increasing from 2328 fans per month to 9065 fans per month, resulting in a total of 71,000 fans by the end of March 2012.

It is observed that the update that went up on the page "Valentine's Day Celebrations at XYZ Hypermarket" on 15 February saw about 544 likes on the post, eventually adding to the increase in fans on the page. Likes help with fan acquisition and with engaging content.

User Demographics

1. The activity on the page has seen an increase in likes through the demographics of the region.
2. The graph represents that most of the activity on the page emanates from the male population, over 65% of the total community on the page.
3. It is seen that the demographics also vary in age group.
4. The age group 25–34 has seen most interaction and engagement on the page.

The "likes" have steadily increased, which indicates that our aim of increasing the reach has been achieved. Likes alone provide very little information. Facebook insights help us to measure the results and response of Facebook activity, namely, number of posts, number of people actively engaging through comments and likes.

Impressions are measured using the edge rank algorithm.

The first metric used is *weekly people talking about this*. This is a measure of unique users sharing stories about the page. These stories include liking the page, posting to the page's wall, commenting on, or sharing one of the page posts, answering a question that was posted, RSVP'ing to one of the events, mentioning the page, photo tagging the page or checking in at your place. This helps to measure the numbers throughout the month. The number is counted only once no matter how many times they interact with the page as only unique users are considered.

The second metric used is "weekly total reach", which is the number of people who have seen any content associated with the page. An important observation here is that if the numbers are falling, then we need to work on the posting strategy and come up with more engaging content. Otherwise, we have to run advertisements/promoted posts to reach more people.

The third metric considered is "virility", which is the number of unique people who have created a story from your page post as a percentage of the number of unique people who have seen it. This needs to be calculated using the formula $100 * (LifetimeTalkingAboutThis/LifetimePostTotalReach)$.

This metric is an indicator of how likely it is that the fans share the posts. The result found is 2.01, which is above the threshold of 1.92 treated as the ideal virility percentage by the edge checker [6]. Facebook focuses on the user experience and integrity of its website. With the increasing number of members spending online, social media advertising is the wave of the future. Facebook seems to have a handle on this new method of advertising that appeals directly to the Internet generation. Facebook advertising and advertisements such as sponsored "like" stories play a very important role in fan increase.

6.4   Case 2: Brand Promotion of a New Cinema Multiplex

In this case, we analyse the problem of brand promotion using social media with Facebook of a new multiplex near Cochin. For the sake of anonymity, the brand is termed ABC cinemas. Data are collected and sourced from an agency contracted by the company. This study examines the growing significance of social media as a tool in achieving the marketing communication objectives of organisations in the emerging economies, in particular, confined to a geographical location. These data are only for a month and are shown in Fig. 2, which is the normalised people's response over the month.

6.5   Case 3: Brand Promotion of a Movie

This is a case study of promoting a movie in Cochin. For the purposes of anonymity, the name is XYZ. Data have been collected via an authorised marketing

Fig. 2 Increase in the number of adopters (in a short span of 1 month)

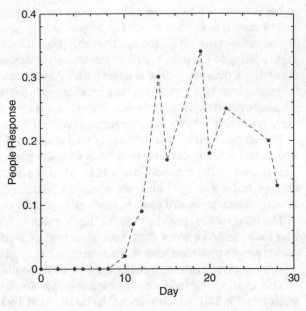

Fig. 3 Daily viral reach of the page by story type. (Unique users) for the first month

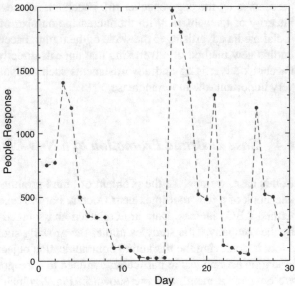

agency. This helped the researcher to become familiarised in the movie industry with creating aggressive viral campaigns on social networks to build awareness, leveraging social media to connect with fans and promoting its product, now the movie. The graph in Fig. 3 depicts the methods of movie marketing and promotion through social media.

6.6 Discussion

In the above three cases studied, the first two follow a similar pattern, whereas the third case gives a totally different picture. The Bass model is a widely adopted diffusion model, used for forecasting the first purchase of a product. The case studies discussed in this chapter are studied within the framework of the Bass diffusion model. The first two almost fit the Bass diffusion model, whereas the third case study is slightly different. It is concluded that domain is also important. In this analysis, we have not considered domicile, gender or age factors. There may be visible changes in the diffusion pattern if these parameters are also considered.

Facebook insight tools [19] provide the metrics that help the study of patterns. In the given cases, it is seen that the "likes" have steadily increased, which shows that our purpose of increasing the reach has been achieved. Likes alone provide very little information. Facebook insights also measure the number of posts, and the number of people actively engaging through comments and likes.

"Weekly people talking about this" is a measure of unique users sharing or making comments on the page. This includes "likes", posting on the page wall, liking, commenting, or sharing any page posts acknowledging events, photo tagging, or location mapping. The number is counted only once no matter how many times they interact with the page as only unique users are taken into consideration.

Weekly total reach is a measure of the number of people who have seen any content associated with the page. An important observation here is that there is a time-related variation. Fans need to be engaged to attract more people.

Virality is the measure of the number of unique people who have created a story from your page post as a percentage of the number of unique people who have seen it. Facebook gives importance to user experience and the integrity of its website. With members spending more time online, social media brand promotion is the future. Facebook seems to be focused on brand promotion that appeals more to the younger generation. Facebook advertising and advertisements such as sponsored "like" stories play a very important role in the fan increase.

The problem studied here is the brand promotion using social media. Social influence exists. We see those patterns in data flows on social networks. There are also visible changes when there is a domain change. The outcome of this social media promotion is:

1. Engaging the fans
2. Improving the fan counts, attracting more fans
3. Brand promotion through social media

Facebook insights provide multiple useful data. In this experiment, we have taken only the three most critical elements to study the reach. Demographics are not considered here. As data collected are confined to a geographical location, other attributes highlighting the local specialities are not considered either.

6.7 Dataset

The dataset analysed here is shared with the understanding of maintaining anonymity. This is used only for academic purposes for methodological theoretical studies. The authors acknowledge the help received in sharing and analysing the real data for brand promotion.

7 Conclusion and Future Research

From the study and discussion of the three case studies presented, the influential aspect of social media in brand promotion is proven, and conforms to the existing Bass model. The Bass model was developed to find out the brand adoption through conventional media. In the study, we analysed the brand promotion activities through social media and the results are very encouraging. In all three cases, it was found that promotion is done across multiple channels and social media initiatives are taken in a long-term sales promotion mode. It was found that major activity of social promotion is within the community, whereas there is much scope for innovation. The intercommunication among the customers, if included, will provide better insights into information diffusion, which has many practical applications. Every attribute has some influence parameter that can be studied in detail in future work.

The data provided by Facebook insights have many attributes such as demographics and geographical frequency of interaction. Variation of any of these attributes may change the behavioural pattern. It is generally found that online consumers are young, working in a company, have a monthly salary and have an account on the social network. Before purchasing a product on online they collect information and feedback from forums, company websites, Facebook accounts or peer reviews.

The study compares the traditional information diffusion model with the new social media model and finds that it conforms with the established patterns. There is much scope for improvement in varying the parameters and modelling the evolving new pattern.

References

1. Srivastava, J., et al.: Data mining based social network analysis from online behaviour. N.p., Web. 2 Aug. 2012, (2008)
2. Osborne, N.: Information diffusion through social networks. http://sikaman.dyndns.org:8888/honours/nicholas-osborne-winter-2011.pdf (2011)
3. Living the Game. http://www.friendster.com (2008)
4. Featured Content on Myspace. http://myspace.com (2009)

5. Ulanat, M. (ed.): Social networks: paradigm shift from information age to connected age. CSI Communications, April 1: 4–6, (2010)
6. Milgram, S.: The small world problem. Psychol. Today **2**, 60–67 (1967)
7. Facebook log in or sign up. http://www.facebook.com (2010)
8. Twitter. https://twitter.com(11 Nov 2012)
9. Linkedin. https://www.linkedin.com (22 June 2013)
10. Haythornthwaite, C: Social network methods and measures for examining e-learning, Social Networks (2005)
11. Wasserman, S., Faust, K.: Social Network Analysis: Methods and Applications, Cambridge University Press, Cambridge (1994)
12. Social Media Marketing. (n.d.) http://www.genacom.com/beesocialmarketing.pdf (2014)
13. Key Facts—Facebook's latest news, announcement and media. http://newsroom.fb.com/Key-Facts (2013)
14. Kincaid, J.: Advocacy groups poke more holes. In: Facebook Privacy, Facebook http://techcrunch.com/2010/06/16/advocacy-groups-poke-more-holes-in-facebook-privacyfacebook-responds (2010)
15. Using social media tools to influence homeowners' building product choices. L.E.K. Consulting, Executive Insights Vol. XIII: Issue 13 (2011)
16. Divol, R., Edelman, D., Sarrazin, H.: Demystifying social media: the McKinsey quarterly. http://www.mckinsey.com/insights/marketing-sales/demystifying-social-media (2012)
17. Bass, F.: A new product growth model for consumer durables. Manag. Sci. **15**(5), 215–227 (1969)
18. Mini, U., Jacob, P.: Information diffusion through social media—analysing brand promotion through Social Network Analysis tools—a business case study. In: Conference on IT in Business, Industry and Government (CSIBIG) (2014)
19. Facebook insights provides data to analyze your business page. http://www.dummies.com/how-to/content/facebook-insights-provides-data-to-analyze-your-b.html

Product Diffusion Pattern Analysis Model Based on User's Review of E-Commerce Application

Niyati Aggrawal, Anuja Arora, Ankit Jain, and Dharmesh Rathor

Abstract Online E-commerce systems such as Amazon facilitate users to provide users' reviews and ratings for products through their daily interaction. The analysis of this social e-commerce giant Amazon shows the presence of a large amount of users' feedback through reviews and its associated features for the published items/products. In this research work, we focused our research on the study of product diffusion pattern analysis as an impact of users' reviews on social e-commerce giant Amazon as a function of time, and to achieve this by introducing a users' review-based rise and fall product diffusion pattern analysis model (PDPA model). The user review-based PDPA model extracts reviews and its associated properties such as review rating, review comment and helpfulness of review using Amazon's application programming interface (API) and assign a weight to all reviews according to the mentioned associated properties. The PDPA model forms graphs according to users' assigned ranking and according to the measured review weight ranking as a time function. These formed review graphs and rating graphs show the diffusion (rise/fall) pattern of product. The review graph further maps the product features and its linked properties to reflect the reason/feature behind the product rise or product fall and measures the review sentiment to show that the diffusion social influence is positive or negative, i.e. the product follows the rise/fall pattern because of the positive influence or negative influence. As a result, we can predict the long-term dynamics of the product on these popular user interactive ecommerce sites by analyzing the users' offered reviews and ratings on these sites. The introduced users' review feature-based rise and fall product diffusion pattern analysis model adds a sense of quality assurance for products and services, which is a new dimension of organic marketing. The conducted experiments confirmed the usefulness of the proposed PDPA model.

Keywords Amazon • E-commerce • Information cascade • Information diffusion • Sentiment analysis • Social influence • Vitality score

N. Aggrawal (✉) • A. Arora • A. Jain • D. Rathor
Jaypee Institute of Information Technology, Noida, India
e-mail: niyati.aggrawal@gmail.com; anuja.arora29@gmail.com; ankit210992@gmail.com; dharmesh17rathor@gmail.com

© Springer International Publishing AG 2017
H. Banati et al. (eds.), *Hybrid Intelligence for Social Networks*,
DOI 10.1007/978-3-319-65139-2_10

227

1 Introduction

The vast amount of shopping done through e-commerce sites and mobile apps has led to an extraordinary increase in e-commerce over the last few years. Shopping has become a convenient, reliable and fast process for users. Although any product can be purchased online today, it has become increasingly difficult for customers to make their purchasing decisions based on product pictures and short descriptions of the product. As many e-commerce sites facilitate users to publicize reviews/feedback of the product on these sites, and these reviews are the diverse, reliable resource for aiding customers, a review is a natural language dialog/conversation and it plays the most important role in online sales. We as human beings evaluate the product quality based on others' perceptions on a particular product in their own natural language. Thus, among the most prevalent features provided by these online media, user reviews and ratings can be efficiently explored to give an in-depth analysis of more relevant opinions to the community [2, 21]. This analysis has a wide range of applications in understanding and predicting the behaviour of the product in the long term [16, 22, 25]. Few e-commerce sites provide an overall rating summary of reviews according to ratings provided by users, as shown in Fig. 1, which is a snapshot of a iPhone6S-Rose Gold review rating summary on Amazon.

On the other hand, a user has to go through all reviews if he/she desires to know about users' reactions for product in detail. However, the number of users' reviews has increased to such an extent where it is no longer possible for a user to peruse them all manually. For example, a digital camera on Amazon has several hundreds of reviews containing thousands of words. To the best of our knowledge, e-commerce websites do not provide product feature summary reports based on users' reviews.

In this research work, we investigate the rise and fall diffusion pattern of a product on the basis of all users' posted evidence: reviews, ratings, helpfulness and time stamp (see Fig. 2) on Amazon e-commerce application using the projected

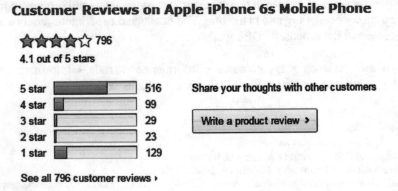

Fig. 1 Review rating summary of iPhone6S-Rose Gold on Amazon

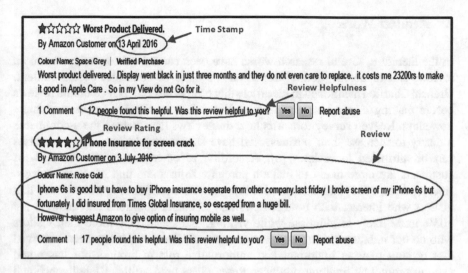

Fig. 2 Users posted evidence on Amazon

product diffusion pattern analysis model (PDPA) as a function of time. This novel approach would be the granularity of the classification of the review datasets, i.e. to show the product rise and fall graph according to the usefulness of the review content (text) as a time stamp. This proposed approach also finds out the promising product features, i.e. those producing the rise/fall pattern of the product and further classifying product feature reviews in positive and negative responses to identifying features that are a prime source of the positive and negative influence of a product on users. This proposed PDPA model may solve a few research domains' problems and can be used to model/predict an activity, such as number of blog postings, people buying products, computer viruses infecting machines, rumours spreading over Twitter and breaking news at a given time stamp.

The rest of the paper is organized as follows. Section 2 summarizes the studied research literature that motivated us to work in this direction. Section 3 presents the research contribution towards the review-based product diffusion analysis direction. Section 4 lays the foundation of the proposed product diffusion-based pattern analysis model. In Sect. 4, relevant definitions, data models and resolved research challenges have been detailed. Review vitality score computation and associated diffusion pattern formation experiments are detailed in Sects. 5 and 6, which discuss how the model is defined and analyzed the review-based rise and fall pattern of product diffusion. Further, Sect. 7 is about how the model identified the promising features of the product that influence users more towards a positive/negative direction Experimental results are demonstrated in Sect. 8 and finally Sect. 9 concludes this research work.

2   Related Work

In the literature, several research works have been carried out to analyze opinion, reviews and to obtain product trends on the basis of consumer-posted content. Graham Charlton mentioned in his article that 61% of customers read online reviews before making a purchase decision, and that is essential for online customers nowadays. Reevoo survey forecasts how the reviews have helped the e-commerce industry to increase their business and have shown that 18% of sales increases can be attributed to reviews [8]. According to other survey reports, 63% of customers are more likely to make a purchase from a site that has user reviews. (iPerceptions, 2011) and Bazaarvoice Site conversion index report mentioned that visitors who interact with both reviews and customer questions and answers are 105% more likely to purchase while visiting, and spend 11% more than visitors who do not interact with user-generated content. Therefore, it has been observed that reviews play an important and authoritative role in product purchases and even in promoting trending products. Researchers used artificial intelligence and text mining techniques to solve the opinion-mining problem of identifying product trends or consumers' opinions for a product from consumers' reviews [4, 15, 19]. On the other hand, sentiment analysis research work is focused on the extraction of the relevance of the product based on the sentiments of consumer reviews expressed in review sentences [11, 12, 23, 24]. Natural language processing [11], supervised learning [24], unsupervised learning techniques [13] and association rules [18, 23] have also been used in sentiment analysis. The sentiment classification relies on the classification of the reviews based on their polarity (positive or negative). Text mining and mutual information are used in sentiment classification. Previous existing approaches have analyze reviews to extract the product's features and classify opinions [4, 11, 12, 18, 23, 24].

In 2004, Hu Minqing et al. [12] mentioned in his research work on "mining and summarizing customer reviews" that as e-commerce becomes increasingly popular, the number of customer reviews that a product receives grows rapidly. For a popular product, the number of reviews may be in the hundreds or even thousands. This makes it difficult for a potential customers to read them to make an informed decision on whether to purchase the product. It also makes it difficult for the manufacturer of the product to keep track and to manage customer opinions. In their work, the authors proposed an approach that mines and summarizes all the customer reviews of a product and actually, the proposed summarization task is different from traditional text summarization because only a product feature summary is required, which the customers have expressed in their review and whether the review is positive or negative. We also integrated this same literature into our work and tried to map a product diffusion pattern with the feature. Graesser et al. created a learning environment in which a conversation is held between mentor and student in natural language. It helps the students to find the solution to a problem that is posed during the conversation between tutor and students. This environment gives students a satisfactory response to their problems in written and documented form with a positive, neutral or negative form [10]. Zhang et al.'s [26] is partially based on and closely related to opinion mining and sentence sentiment analysis classification. Extensive research has been done on sentiment analysis of review text and subjectivity analysis (determining whether a sentence is subjective or objective). Another related area is feature-/topic-based sentiment analysis, in which opinions on particular attributes of a product are determined. Most of this work concentrates on finding the sentiment associated with a sentence (and in some cases, the entire review). There has also been some research into automatically extracting product features from review text. Although there has been some work on review summarization and assigning summary scores to products based on customer reviews, relatively little has been done on ranking products using customer reviews [26]. To the best of our knowledge, there have been no focused studies on product feature-based ranking using customer reviews. The most relevant work is where the authors introduce an unsupervised information extraction system that mines reviews to build a model of important product features, and incorporates reviewer sentiment analysis to measure the relative quality of products. Later, in 2014, Zhang et al. used natural language processing techniques to identify sentences in reviews that provide subjective and comparative information regarding products. The goal of this work is to help customers make better decisions without having to read a large repository of customer reviews. Their experimental results indicate that their product ranking is consistent with rankings done by subject experts [27]. Kim, Young, and Jaideep Srivastava discussed the impact of social influence on the decision regarding the sale of online products. The authors discussed the important issues that have the greatest impact on the sale of online products. These major issues confine the social communication about the product to e-commerce sites, to merge user choices regarding the product with social pressure and to work out the social influence on customers' purchase decision-making, so as to expect the greatest impact of social influence in e-commerce [14].

Matsubara et al. studied the rise-and-fall patterns in the information diffusion process through online media [17] and presented a SpikeM model to model/predict an activity (e.g. number of blog postings, people buying products, computer viruses infecting machines, rumours spreading on Twitter etc.) as a function of time [1]. The SpikeM model was useful for "short-term" forecasting and for answering "what-if" scenarios and spotting outliers. Later, in 2014, Bancken, Wouter, Daniele Alfarone and Jesse Davis presented a work on automatically detecting and rating product aspects from textual customer reviews and it matches the behaviour of numerous, diverse, real datasets including the power-law decay [3]. In a case study given by Joyce Chai and Jimmy Lin, the authors concluded that a natural language system is more efficient for obtaining the relevant information about any product or any object than the menu-driven system. Authors proposed a prototype system that provides natural language dialogue capabilities to help users to access e-commerce sites to find relevant information about products and services. The authors' proposed prototype is capable of facilitating technology in NLP and human-computer interaction, which is a more rapid and more instinctive way of interacting with e-commerce sites, especially for inexperienced clients [7].

According to the studied literature, this research work focuses attention towards the context in which reviews are expressed. The proposed approach understands and summarizes posted reviews in lieu of product feature-based context. In this work, we present an approach that is required to identify product contexts/product features consist into the review and determine whether the review expresses the consumer's preference towards the product. The product rise or diffusion pattern is dependent on product features, which can be identified from the posted users' reviews and the influence of a specific feature is also an essential criterion for product preference. Therefore, without such information, suggesting/recommending a preference is of little or no practical use. Thus, one should not only talk about preferences extraction from consumer reviews, but also about the feature information for which preferences have been expressed. This information is easier for a consumer to express in a review, which he can write using free text form. Thus, we need to be able to analyze the natural language text accurately to identify and extract user's preferences and the features on which preferences have been expressed. Thus, we obtain a diffusion index by analyzing and providing results about how the new products are going to diffuse (spread) and what quality of product should be released by a company to increase diffusion with a positive influence.

3 Research Contribution

Because of the rising internet culture, the e-commerce business is booming, with more and more people being attracted to purchasing products from these online portals. Therefore, there is a need to analyze customer reviews. With thousands of reviews for every product, it is necessary to summarize and efficiently showcase these reviews to make them useful. There are basically two contributions in this paper: First, we characterize the human communication properties of review content, including human-performed actions on review for determining the review influence on consumers while purchasing/looking for a product. Second, we measure and extract the consequence factors via reviews that are having an impact, to spread/diffuse product popularity.

To achieve the above-mentioned contributions, there are three steps to our approach:

- We detect and elicit the linguistic keywords/consequence factor (such as product features) of the product under consideration using a sophisticated "term extraction model".
- We determine the review-based diffusion pattern (rise/fall pattern) of the product based on the human communication properties, computed product reviews and the social influence factor using the product diffusion pattern analysis model.

- Finally, we infuse the product diffusion consequence factors, which are the prime reasons for the product diffusion.

4   Foundation of Diffusion Analysis

In this section, we discuss our problem formulation definitions, data model and product diffusion pattern analysis model.

4.1   Definitions

Here, we define the concept involved in this research work to discuss and solve the defined research contributions.

1. Information cascade & diffusion: Information cascade or diffusion occurs when the consumer/user observes the actions of others and further engages in that same act despite their own actions. Each new consumer/user makes a decision based on previous users' adopted views or ideas. We define the diffusion pattern of a product based on the possibility of review adoption, which is represented as a real non-negative scalar dependent on:

   (a) Elicited linguistic keywords that are set with regard to the product features/product properties discussed in the reviews at various time stamps
   (b) Comments on that review
   (c) Helpfulness measure of that review

2. Social influence: This is the result of rational inference (positive/negative) from limited information. The buyer/consumer/user makes decisions sequentially and the actions of previous consumers affect those of later consumers. We propose that the social influence of a review can be characterized by:

   (a) Elicited product feature/product properties
   (b) Its associated influence (positive/negative)
   (c) "Helpfulness" count
   (d) "Conversational comment" count The helpfulness and conversational comment count assign weightage to the negative and positive social influence.

3. Consequence factor: The consequence factors of the salient product features and other product properties associated with conversations in different reviews are set in time.

4. Conversational comment: The conversation comment count of a review shows the influence and weightage of product features/product properties associated with that conversational comment.

5. Helpfulness: Amazon offers a feature whereby the consumer/user can mark the helpfulness of a review and count the helpfulness of a review showing its importance and the social influence of the product.

4.2 Data Model

The data model used comprises the tuple $\{R_E, R_A, T_R, U_R, C_R\}$ with the following attributes of the product under consideration: collection of reviews R_E; collection of review ratings R_A of the product; collection of time stamps of the reviews T_R when the review is posted; collection of users posting reviews U_R. Each review further contains a set of comments C_R, such that each comment that belongs to a specific review is associated with that review; comment content; helpfulness measure, which shows the usability of the review to consumers/buyer and the time stamp.

We assume that "u" users posted R_E review at time R_T with R_A rating, H_R helpfulness measure and C_R number of comments for the R_E review. K is a set of the total number of consequence factors (such as product features and other product-specific properties) discussed in all the reviews.

4.3 Research Challenges

The formal definition of our problem statement to achieve the above-mentioned research contribution is mentioned here. We intend to determine the product diffusion pattern, its social influence and infuse the consequence factor with the help of a extracted review-related data set $\{R_E, R_A, T_R, U_R, C_R\}$ and review content. The idea is to leverage the capillary reviews to be able to retrieve the diffusion pattern and consequence factors out of the terms used in the reviews.

The following challenges are raised to justify the research contributions via this attempt:

RC1: How to elicit the consequence factors/linguistic keyword from the reviews and comments.

RC2: How to model the diffusion (rise & fall) pattern of a product based on the consequence factors mentioned in that review and review-associated properties.

RC3: How to achieve the social influence (negative/positive/neutral) from a review and its associated comments.

RC4: How to infuse the consequence factor, which is the main source of social influence and diffusion pattern of a product?

In Sects. 5 and 6, we mention the method of addressing and resolving the above-mentioned research challenges.

5   Content Extractions

The diffusion analysis process starts with the extraction of content, such as predefined terms and the consequence factor from the product page on Amazon and from the stream of reviews. Thereafter, we extract all reviews and its associated content from the Amazon e-commerce site. In particular, we extract the following content (Fig. 3):

5.1   Term Extraction Model

We elicit all product features, product properties and aggregate collections of predefined terms. Predefined terms are the most probable technical terms used in reviews of the product under consideration and these predefined terms are indirectly origin or prime consequence factor for a rise/fall diffusion pattern of a product. To do this, we used the aggregate model to collect all probable terms that users usually use when posting a product review.

In this aggregate model, we extract the corpus "PT", with n= |PT| predefined terms extracted from the product page on Amazon using the HTML DOM parser library, and to enrich the term corpus PT, we exponentially add terms K also used in the product reviews. We used NLTK, an open source library [20] to extract terms from the review. First, all the URLs and punctuation are removed from the reviews. Second, the less relevant words, also called the stop words, are removed from the reviews. Thereafter, the stemming technique called the Porter stemmer is applied to convert the words into stems (root words). This is all easily achieved through the process of NLP. Review filtering helps to remove unnecessary information. The final keywords obtained after the pre-processing of reviews are stored in PT. Thus, the predefined term set PT is the collection of terms extracted from the Amazon product page and terms extracted from product reviews.

Fig. 3 Predefined term set

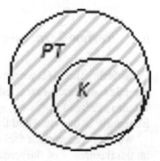

5.2   Review and Associated Content Extraction

As already mentioned in the data model section, we extract the corpus RE, with m=$\|RE\|$ reviews and we associate a representative review vector re_i with each review re_i, which represents the associated content of the review. We preserve all the review-associated content, such as the review's posted time stamp rt_i, the review rating ra_i, review helpfulness hr_i and the number of comments on the review cr_i. Thus, for each review re_i, a review vector is defined as $re_i = \{rt_i, ra_i, hr_i, cr_i\}$. After this, all the information related to reviews has been formally defined as a review vector.

Further, we observed that a review contains a number of terms and even the same term is available a number of times in a review. Another observation is that users can give a reaction to a particular review in the form of a comment on the Amazon commerce application. Therefore, while measuring the review's term weight, we included these two observations as well.

6   Product Reviews Vitality Score and Diffusion Pattern

In this section, we present the statistical analysis measure of the vitality of the review to measure the role of reviews in making a product popular or unpopular using quantitative and qualitative measures of the terms used in the review. Therefore, this section is all about computing the vitality score of each and every review and mapping it to the diffusion pattern of the product under consideration into the Amazon e-commerce application.

To compute the vitality of each review the considered statistical measures are as follows:

Review term vitality score
Review reactions vitality score

6.1   Review Term and Term Vitality Score

Considering the idea, we associate each review re_j with a term vector t_j that contains the relevant terms extracted from that review. Thus, for each review re_j, a term vector is defined as $t_j = \{t_1, t_2, \ldots, t_x - 1, t_x\}$, but we also highlight terms that occur frequently in a review and that are highly relevant for a product.

We bind the weight factor [6, 20] with each term that exists in a review based on the frequency of the term used in the review [6, 20]. Therefore, we calculate the term frequency weight $w_r ei, t_i$ of the t term in re review by using Eq. (1).

$$tfw_{re_i,t_j} = \frac{tf_{re_i,t_j}}{tf_{re,t_j}^{\max}} \tag{1}$$

where tf_{re_i,t_j} the term frequency of j-th term in i-th is review; $tf_{re,t_j}^{\max}$ is the maximum term frequency of j-th term out of all the reviews of a product. Computed w_{re_i,t_j} is a normalized weight to prevent a bias towards bigger documents, e.g. frequency divided by the maximum frequency of any term within the document [5].

Term frequency has a significant problem in that all terms are considered equally important when it comes to assessing the vitality of the review. In reality, certain terms have little power in characterizing the document. Hence, an inverse document frequency factor is incorporated that diminishes the weight of the terms that occur very frequently in the document set, increases the weight of terms that occur rarely and is computed using Eq. (2).

$$idfw_{re_i,t_j} = \ln \frac{\|re_i\|}{1 + \|re_i t_j\|} \tag{2}$$

where $\|re_i\|$ defines the total number of documents in the corpus and $\|re_i t_j\|$ the number of documents in which the term appears. Thus, the weight of a term t_j in the review re_i is defined as

$$w_{re_i,t_j} = tfw_{re_i,t_j} \times idfw_{re_i,t_j} \tag{3}$$

Finally, for review re_i the term weight vector tw_{re_i} comprises:

$tw_{re_i} = \{w_{re_i,t_1}, w_{re_i,t_2}, \ldots \ldots \ldots, w_{re_i,t_j}\}$

In this process, all the terms or product information expressed in the review have been defined as the review term weight vector.

Thereafter, as already mentioned in Sect. 5, all the user reviews and associated comments under a particular product combine to form a predefined term set/term bag. Consequently, all the term weights corresponding to the existence of terms in the whole review set is calculated using the above-mentioned Eqs. (1), (2), and (3), and stored as w_{B_k,t_j}, which defines the weight of the term t_j in the predefined term set of product k corresponding to term usage in all reviews. Thus, for a specific product k, the predefined term set B_k term weights vector is defined as $tw_{B_k} = \{w_{B_k,t_1}, w_{B_k,t_2}, \ldots \ldots \ldots, w_{B_k,t_j}\}$.

The review term vitality score tvs_{re_i} of each review is calculated as the summation of all review term weight proportions in relation to that same term weight in the predefined term set (term bag) and computed as:

$$tvs_{re_i} = \sum_{t_j=1}^{x} \frac{w_{re_i,t_1}}{w_{B_k,t_j}} \tag{4}$$

where the term vitality score tvs_{re_i} according to the term's presence in a review. It is the summation of each term weight w_{re_i,t_1} in that review in proportion to that same term weight w_{B_k,t_j} in the predefined term set, where x is total number of terms occurring in review i, and k is product for which we are analyzing the diffusion pattern.

6.2 Review Reactions Vitality Score

It has been observed that e-commerce sites facilitate users to post their reactions/feedback on reviews, such as review rating, review helpfulness measures and comment counts. These reactions supplement consumers' behaviour towards the product. Therefore, we compute the review vitality score corresponding to users' reactions, which constitute the main promotion factor of users' perspective of the product. Users can post a reaction to a review in three forms: by assigning a rating to a review, by posting a comment on the review and by providing a helpfulness measure to a review.

6.2.1 Review Rating Vitality Score

Users assign a rating to a product on a scale of 5 on Amazon. Thus, we attach a review rating vitality score $ravs_{re_i}$ with the entire review vector and the $ravs_{re_i}$ score is 0.2, 0.4, 0.6, 0.8, 1 for 1, 2, 3, 4, 5 ratings respectively for all reviews.

6.2.2 Review Helpfulness Vitality Score

Users can acknowledge reviews by mentioning their helpfulness and the number of users who found that review useful is shown on Amazon. For this, we choose a helpful review with a maximum count and find out all the review helpfulness vitality scores $hrvs_{re_i}$ proportionate to the most helpful review. This is calculated as:

$$hrvs_{re_i} = \frac{\|hr_{re_i}\|}{\|\max(hrvs_{re_i})\|} \tag{5}$$

6.2.3 Review Comment Count Vitality Score

A user's reaction in the form of a comment is an important measure. The terms used in the comments are already being considered, whereas the computing term vitality of review, i.e. review terms include associated comment terms too. However, we also consider users' comment counts as a measure of the vitality score. For this, we compute the comment count vitality score $crvs_{re_i}$ as:

$$crvs_{re_i} = \frac{\|cr_{re_i}\|}{\|\max(crvs_{re_i})\|} \tag{6}$$

where $\|cr_{re_i}\|$ is the comment count of a product review for which we are calculating the comment count vitality score and $\|\max(crvs_{re_i})\|$ is the maximum review comment count.

6.2.4 Product Review Vitality Score

In the subsections above, we computed four vitality scores corresponding to user reactions. The Aggregated review vitality score RVS_{re_i} is a summation of all computed vitality scores: term vitality score tvs_{re_i}, review rating vitality score $ravs_{re_i}$, review helpfulness vitality score $hrvs_{re_i}$, and the review comment count vitality score $crvs_{re_i}$.

$$RVS_{re_i} = tvs_{re_i} + ravs_{re_i} + hrvs_{re_i} + crvs_{re_i} \tag{7}$$

All review vitality scores are computed according to their use to customer and to provide an accurate review vitality/weight according to the review content compared with the entire review set of a product.

7 Social Influence Measurement

There is a misconception in the previously defined PDPA process. According to this, the review shows a rise or fall pattern according to the review properties, such as the terms used in the review, the rating, the helpfulness, the comment count and the comments. Assume a case in which a negative review has a higher rating, a higher helpfulness index and more comments, then the product diffusion pattern graph shows a peak, even though the product has a negativediffusion (spread). This means a peak or nadir of a graph shows how helpful the review is in spreading the product's popularity, but it is not certain that the popularity of the product based on reviews is in a negative direction or a positive direction. The review peak represents the review's importance in spreading product popularity.

Therefore, we also measured the influence of the terms used in the review, i.e. the terms/product features mentioned in the review contain a negative/positive influence, and while depicting the diffusion graph, we additionally show the influence of the product, for example, the product diffusion graph shows a peak for a particular review, but that peak/rise is due to a negative influence/positive influence.

To do this, we applied the text edge processing algorithm [9] to determine the frequency of occurrence of predefined terms and concepts in conjunction with the

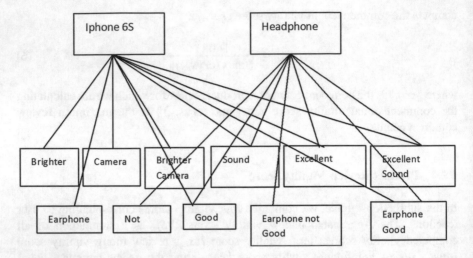

Fig. 4 Text edge processing algorithm example

relevant product features and with the adjectives/adverbs found. In the text edge processing algorithm, the terms relationship has been analyzed to determine the edge relationship strength and its sentiment. The steps are as follows:

- We already parsed the review into individual words using filtering and stemming techniques while ignoring the stop words and words with little information such as "of", "it", "is" etc. are removed, as described in detail in Sect. 5.

Next, the relationship between the main predefined term of interest and each found word or tuple is stored. An example to depict the text edge processing algorithm is shown in Fig. 4.

- Each relationship is then counted as one instance of an "edge" between these connected objects.
- The number of relationships (example: good battery, excellent camera etc.) between objects is added up.
- We used the SentiWordNet interface using the National Language Toolkit corpus in Python to classify the reviews into positive and negative relationships.
- The number of positive and negative relationships defines the positive/negative influence measure. The resulting positive and negative influences are measured according to the frequency of the number of positive/negative relationships in a review.

8 Experimental Results

8.1 Dataset Details

The dataset that we have used for this work consists of reviews from the e-commerce site Amazon.com. The dataset spans a period of approximately 2 years. The dataset of product reviews includes information about the product, user information, review ratings, and a plaintext review. We show experiments using the proposed approach performed on two products: iPhone and headphone datasets obtained from Amazon and, as clearly stated above, our approach is centred around user-generated ratings and their linked comments. Amazon, being one of the largest e-commerce companies, engages a large number of customers all over the globe. This platform provides a rich repository of reviews, ratings and related metadata for analysis from a large base of users. Our motive for using these two datasets is to cover different aspects of product diffusion. Detailed descriptive statistics for the products taken is depicted in Table 1.

8.2 Results: Diffusion Pattern

This approach depicts the diffusion pattern corresponding to user reviews for a specific product, which helps companies to identify any problems with the product and to know what updates are required. It also helps the company to determine when to launch a new product. Experiments on real-world datasets show that our

Table 1 Descriptive statistics for the iPhone andheadphone dataset

iPhone dataset
Number of reviews: 1500
Number of ratings: 1500
Average number of comments in review dataset: 25
Time stamp: January 2013 to September 2015
Average number of terms/words in review: 27
Maximum number of terms in review: 266
Sennheiser HD 202 professional headphones (black) dataset
Number of reviews: 400
Number of ratings: 400
Average number of comments in review dataset: 2
Time stamp: January 2013 to September 2015
Average number of terms/words in review: 35
Maximum number of terms in review: 1718

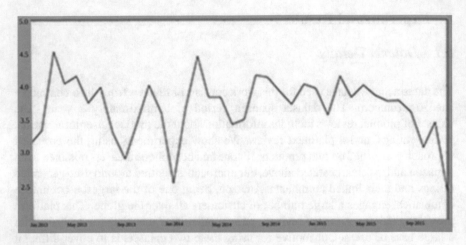

Fig. 5 iPhone user rating pattern as a time function

vitality measurement approach produces results that correspond well with Amazon's manual ranking, which is calculated by subject experts.

Figure 5 depicts the behaviour of the iphone rating graph from January 2013 to September 2015. This graph is simply the depiction of iPhone user ratings obtained for a particular period. It is clearly visible that between January 2013 and March 2013, the rating given by the customers increases and after that it starts to decrease, until September 2013 when it started to increase again; thus, the iPhone has a changing pattern corresponding to the time stamp.

Consumers can undergo a comprehensive shopping experience with the advent of product reviews so that purchases can be made specific to a customer's need and a product's features. Amazon offers a large pool of user reviews that can serve this purpose. We have constructed the dataset using Amazon's product advertising application programming interface (API) for the iPhone and Sennheiser HD 202 professional headphone data for the period January 2013 to September 2015.

Figure 6 depicts a graph in between the time stamp and the computed aggregate review vitality score RVS_{re_i}, which shows the rise and fall pattern of the iPhone according to reviews given by customers. The aggregated review vitality score RVS_{re_i} is a summation of the term vitality score tvs_{re_i}, review rating vitality score $ravs_{re_i}$, review helpfulness vitality score $hrvs_{re_i}$, review comment count vitality score $crvs_{re_i}$. This graph shows when the customers were interested in the product and when they disliked it.

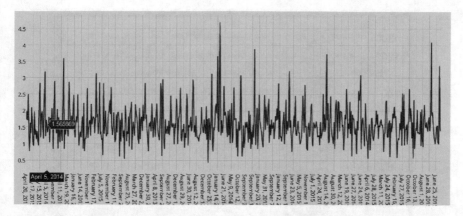

Fig. 6 iPhone diffusion pattern according to the review virality score corresponding to time function

8.3 Results: Social Influence Mapping with Diffusion Pattern Graph

We take into account the review terms and adjectives/adverb words and classify them as neutral/positive/negative. We create the adjective/adverb words database using the dictionary API "Big Huge Labs", which takes adjectives used in reviews and generates synonyms and similar words. We applied the text edge processing algorithm [9] to determine the frequency of occurrence of predefined terms in conjunction with the relevant product features and with the adjectives/adverbs found. To validate the efficiency of the text edge processing algorithm, we applied it to three mobile phone brands, LG, Motorola and Samsung, on three sites, social networking sites Facebook, e-commerce site Flipkart and review analysis site CNET.

Figure 7 shows the results for LG brand phones on the three chosen domain web sites. It shows that observed pattern of the Flipkart dataset is substantially different, not having any negative sentiments for LG products and does not even have high numbers of positive and neutral sentences.

The embodiments illustrated in Fig. 8 are LG brand phones and the graph shown represents the number of relationships between the main term of interest and the word feature in the three domain-extracted sentences. As depicted in Fig. 7, Flipkart does not have negative sentiments; thus, the same results have been achieved using the text edge processing algorithm. The CNET review site and the Flipkart e-commerce site feature-based brand popularity graphs produce almost the same popularity graph. Using the text edge processing algorithm, we are able to analyze feature-based brand popularity.

We designed a function that integrates the diffusion rise and fall review pattern graph of all reviews and displays a word cloud of the selected review to understand consequence factors of the rise and fall pattern and their influence. This function

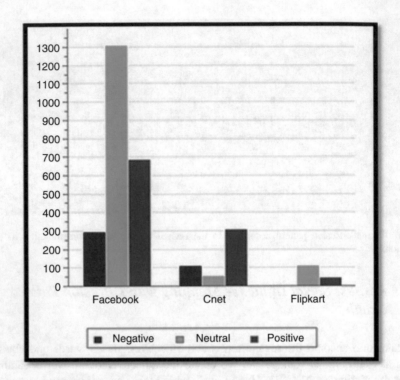

Fig. 7 Sentiment analysis for the LG brand on Facebook, CNET and Flipkart

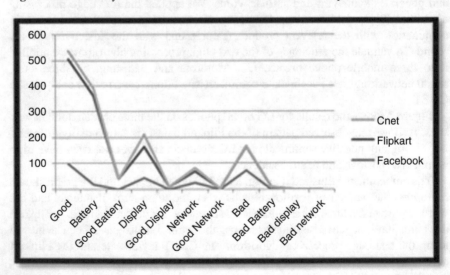

Fig. 8 Text edge processing for LG brand phones

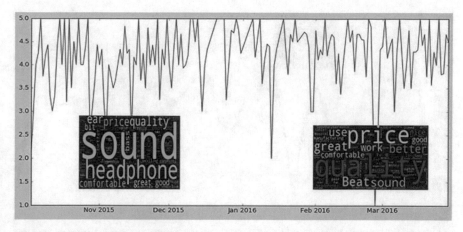

Fig. 9 Diffusion pattern graph of headphones with consequence factor of the rise and fall pattern

shows the word cloud of the terms used and its associated adjectives/adverbs for the number of reviews at a specific selectedtime stamp, as shown in Fig. 9. In word cloud terms, adjective or adverb size shows the frequency of terms available in a specific review set for a specific time stamp. As already mentioned in Sect. 5, we segregate predefined terms set on the basis of features available on the product page on the e-commerce site and integration of the terms used in reviews with the adjectives mentioned. Figure 9 shows a two-word cloud/consequence factor of reviews with a specific time stamp. The first word cloud is of 10 November 2015 review set and shows the fall pattern. This review contains terms such as sound, price, comfort etc., but this review has not gained many users' attention. On the other hand, the second word cloud of reviews of 23 February 2016 has a peak, which means that people are influenced by these reviews and the consequence factors/terms of this review set are quality, price, work etc., which clearly shows the overall features and quality of the product, providing the diffusion pattern with a peak, whereas previous reviews were much towards sound and price. Therefore, the reasons behind the popularity spread in the second review are all of its features.

This diffusion pattern rise or fall shows the influence of a review on users, but whether the influence is positive or negative is again a prime measuring factor. Social influence towards the product can be seen through the peak of the diffusion pattern graph. Peak represents whether the social influence is positive or negative. Through this, we also see the popularity of the product popularity is down corresponding to which features. To do this, we again used the text edge processing algorithm and the text edge processing algorithm measure sentiment analysis of the review.

Finally, the social influence within the diffusion graph, as shown in Fig. 10 for the iPhone dataset, are integrated and represented. The area visualized in red shows negative reviews and the computed review vitality score is negative, whereas the area shown in blue shows a positive influence. Figure 10 shows a diffusion pattern

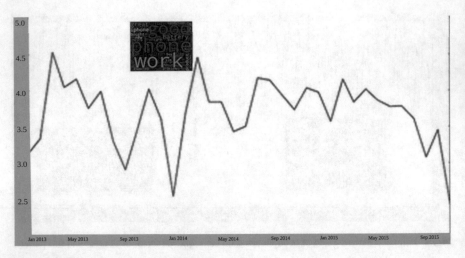

Fig. 10 Diffusion pattern graph of the iPhone corresponding to the social influence and affected consequence factors of the rise and fall pattern

graph for the iPhone corresponding to the reviews' social influence and consequence factors for the affected product features or product quality as well.

9 Conclusions

In this research work, we have fully created a social network application-independent review-based product diffusion pattern analysis model. This model works very well for all the review-based systems, as it has been tested on other review-based systems and on, for example, video users' comments on YouTube.

The main contribution of this article consists of the successful design of the review diffusion pattern and the further mapping of it with the consequence factors of the rise and fall diffusion and with positive or negative influence measures.

In this process, the peak of product diffusion graph shows the interestedness of users in that particular review, but this peak is associated with an influence on users that is either positive or negative. This influence has been achieved using the text edge processing algorithm to measure review sentiment. Further, product features discussed in the review depict the consequence factor behind the positive and negative influences. The proposed approach has few shortcomings, such as a particular review having positive and negative reviews, but the system computes an overall sentiment of that particular review and assigns weight to terms used in a specific review based on its occurrence in that review. In future, we may further improve the approach by extracting the consequence factor of a review, at the same time as displaying the consequence factor together with its influence.

References

1. Aggrawal, N., Arora, A.: Vulnerabilities issues and melioration plans for online social network over Web 2.0. Commun. Dependability Qual. Manag. Int. J. **19**(1), 66–73 (2016)
2. Aggrawal, N., Arora, A.: Visualization, analysis and structural pattern infusion of DBLP co-authorship network using Gephi. In: 2016 2nd International Conference on Next Generation Computing Technologies (NGCT). IEEE, Piscataway (2016)
3. Bancken, W., Alfarone, D., Davis, J.: Automatically detecting and rating product aspects from textual customer reviews. In: Proceedings of the 1st International Workshop on Interactions Between Data Mining and Natural Language Processing at ECML/PKDD, pp. 1–16 (2014)
4. Behl, D., Handa, S., Arora, A.: A bug Mining tool to identify and analyze security bugs using Naive Bayes and TF-IDF. In: 2014 International Conference on Optimization, Reliability, and Information Technology (ICROIT), pp. 294–299. IEEE, Piscataway (2014)
5. Buche, A., Chandak, M.B., Zadgaonkar, A.: Opinion mining and analysis: a survey. arXiv:1307.3336 (2013, arXiv preprint)
6. Cataldi, M., Di Caro, L., Schifanella, C.: Emerging topic detection on twitter based on temporal and social terms evaluation. In: Proceedings of the 10th International Workshop on Multimedia Data Mining. ACM, New York (2010)
7. Chai, J. et al.: The role of a natural language conversational interface in online sales: a case study. Int. J. Speech Technol. **4**(3–4), 285–295 (2001)
8. Charlton, G.: Ecommerce consumer reviews: why you need them and how to use them. Econsultancy.com (2012)
9. Goeldi, A.: Website network and advertisement analysis using analytic measurement of online social media content. U.S. Patent No. 7974983 (5 July 2011)
10. Graesser, A.C. et al.: AutoTutor: a tutor with dialogue in natural language. Behav. Res. Methods Instrum. Comput. **36**(2), 180–192 (2004)
11. Haddi, E., Liu, X., Shi, Y.: The role of text pre-processing in sentiment analysis. Procedia Comput. Sci. **17**, 26–32 (2013)
12. Hu, M., Liu, B.: Mining and summarizing customer reviews. In: Proceedings of the 10th ACM SIGKDD International Conference on Knowledge Discovery and Data Mining, pp. 168–177. ACM, New York (2004)
13. Katta, E., Arora, A.: An improved approach to English-Hindi based cross language information retrieval system. In: 2015 18th International Conference on Contemporary Computing (IC3), pp. 354–359. IEEE, Piscataway (2015)
14. Kim, Y., Srivastava, J.: Impact of social influence in e-commerce decision making. In: Proceedings of the 9th International Conference on Electronic Commerce. ACM, New York (2007)
15. Kushal, D., Lawrence, S., Pennock, D.M.: Mining the peanut gallery: opinion extraction and semantic classification of product reviews. In: Proceedings of the 12th International Conference on World Wide Web. ACM, New York (2003)
16. Li, Q., Wang, J., Chen, Y.P., Lin, Z.: User comments for news recommendation in forum-based social media. Inf. Sci. **180**(24), 4929–4939 (2010)
17. Matsubara, Y., Sakurai, Y., Prakash, B.A., Li, L., Faloutsos, C.: Rise and fall patterns of information diffusion: model and implications. In: Proceedings of the 18th ACM SIGKDD international conference on knowledge discovery and data mining, pp. 6–14. ACM, New York (2012)
18. Ngai, E.W., Xiu, L., Chau, D.C.: Application of data mining techniques in customer relationship management: a literature review and classification. Expert Syst. Appl. **36**(2): 2592–2602 (2009)
19. Pang, B., Lee, L.: Opinion mining and sentiment analysis. Found. Trends Inf. Retr. **2**(1–2), 1–135 (2008)
20. Salton, G., Buckley, C.: Term-weighting approaches in automatic text retrieval. In: Information Processing and Management, pp. 513–523 (1988)

21. Siersdorfer, S., Chelaru, S., Pedro, J.S., Altingovde, I.S., Nejdl, W.: Analyzing and mining comments and comment ratings on the social web. ACM Trans. Web (TWEB) **8**(3), 17 (2014)
22. Szabo, G., Huberman, B.A.: Predicting the popularity of online content. Commun. ACM **53**(8) 80–88 (2010)
23. Tang, H., Tan, S., Cheng, X.: A survey on sentiment detection of reviews. Expert Syst. Appl. **36**(7), 10760–10773 (2009)
24. Ye, Q., Zhang, Z., Law, R.: Sentiment classification of online reviews to travel destinations by supervised machine learning approaches. Expert Syst. Appl. **36**(3), 6527–6535 (2009)
25. Yee, W.G., Yates, A., Liu, S., Frieder, O.: Are web user comments useful for search. In: Proceedings of LSDS-IR, pp. 63–70 (2009)
26. Zhang, K., Narayanan, R., Choudhary, A.N.: Voice of the customers: mining online customer reviews for product feature-based ranking. In: WOSN, vol. 10, p. 11 (2010)
27. Zhang, K., Narayanan, R., Choudhary, A.: CUCIS technical report mining online customer reviews for ranking products (2014)

Hierarchical Sentiment Analysis Model for Automatic Review Classification for E-commerce Users

Jagbir Kaur and Meenakshi Bansal

Abstract The trend of online shopping has grown over recent years. Assessing the quality of the product has become very important. Product review classification is used to analyze the sentiment from reviews posted by the user to prepare the product report. In this chapter, the mechanism has been proposed for opinion mining over text review data for the generation of product review reports based upon multiple features all together. This report shows positive and negative points about the specific product, which can play a significant role in the selection of the products on online portals. The proposed model has been evaluated based on the performance parameters of the precision, recall and polarity-based accuracy assessment, which gives the overall perspective of the overall accuracy. The proposed model has been clearly defined as being better than the existing models, when assessed on the given parameters.

Keywords Aspect-based classification • Opinion mining • Polarity • Polarization • Product reviews • Word stemming

1 Introduction

Product reviews play a vital role in a selection of a particular product. Customer reviews about a product are considered to be sales drivers and are something that the majority of the customers will want to know before making a decision about buying a product [13]. It is a fact that online customer reviews are trusted nearly 12 times more frequently than the description provided by the manufacturers. Inventors of e-commerce such as Amazon and eBay have been using product reviews since 1997; they lead people to write their opinion and share their experience of the products they have used.

J. Kaur (✉) • M. Bansal
YCoE, Punjabi University, Guru Kashi Campus, Talwandi Sabo, India
e-mail: bhakerj@gmail.com; ermeenu10@gmail.com

© Springer International Publishing AG 2017   249
H. Banati et al. (eds.), *Hybrid Intelligence for Social Networks*,
DOI 10.1007/978-3-319-65139-2_11

1.1 Need for Customer Reviews

Gathering reviews from customers act as an asset for an organization selling products, as this helps manufacturers to be aware of the strengths and weaknesses of their product and help them to improve it. When going to buy a product online, customers usually look at the product's ratings, read reviews given by other customers and then compare the product with other products of the same category. Quite simply, customer reviews increase conversions. Customer reviews help to improve online business. Organizations look out for the reviews given by customers to know what improvements they can make to their product. Different organizations use several different ways of obtaining product reviews from customers. For instance, Amazon has a brilliant model for gaining product reviews through email. Social media such as Facebook, Twitter and many others are considered reliable sources of getting reviews. Using customer service or suggestion cards, customers are suggested to leave their thoughts and opinions about products. However, it is difficult for a customer to go through hundreds or thousands of reviews to make a decision whether to buy a product or not. In response, this chapter has described a technique for the summarization of customer reviews [16]. There are various factors that show the importance of customer reviews for an organization selling products online:

1. Whenever a company introduces a new product, customer feedback is very important for determining the customer's needs and tastes.
2. Companies can better understand how their products are better than competitive products by analyzing the customer ratings of the product and their reasons for selecting it.
3. Companies can determine whether their customers are getting a satisfactory level of service from their employees.
4. Customer reviews help to determine why consumers are no longer interested in buying products from them, if any. This will assist with building up strategies that would help lure customers back into the business.
5. Customer reviews are also important in determining technological trends in the market.

1.2 Processing of Online Client Reviews

Early works aimed at classifying entire documents as containing overall positive or negative polarity, or rating scores of reviews. Most of the existing methods of processing online customer reviews focus on opinion mining, which is aimed at discovering whether a reviewer's attitude is positive or negative with regard to the various features of the product. For an example shown below as example 1, opinions regarding the picture quality of a digital camera are specified [1].

- Digital Camera
- Feature: Image Quality
- Positive: three hundred (review statements)
- Negative: fifty (review statements)

Example 1: An Example of Opinion Mining

In the above example, the picture quality of a digital camera is considered to be good because more opinions are positive. However, there are some customer comments or reviews that cannot be tagged as positive or negative, but that are still useful. For example: Reviews regarding a Nokia phone started as:

- The phone's sound quality is great.
- The most important thing for me is sound quality.

Here, both the sentences are talking about product reviews featuring sound quality. But the second sentence does not interpret any attitude orientation that is either positive or negative. However, it expresses the customer liking that what the customer wants from the product is sound quality. It is equally important to know that this review cannot be ignored. Such methods are not considered in opinion mining. Moreover, opinion mining focuses only on product features, but product features cannot cover all significant issues in customerreviews. For example, most of the customers, when talking about the Nokia 6610 phone, they have written about the flip phone. These reviews are critical to understanding the rationale of decision-making and purchasing. But opinion mining does not consider the flip phone to be a feature of the product. In previous works, opinion mining or sentiment analysis was mainly focused on extracting the opinions from customer reviews. Opinion mining predominantly determines whether the review sentences convey a positive, negative or neutral orientation. However, it is not enough to study only positivity or negativity in reviews. In other words, concentrating only on positive or negative comments is not enough to cover all the important topics and different concerns across different reviews. Therefore, this work focuses on a technique that generates a summary based on the essential topics mentioned in the reviews [13]. Thus, a summary obtained helps customers to take a decision about buying a product or not. In this internet age, many different techniques are used today to predict appropriate evaluation regarding which product to buy by drawing conclusions from all the reviews of a product given by different consumers. Such techniques include feature-based summarization, summarization by fuzzy logic, summarization through lexical chains, frequent pattern mining algorithm and many more. This chapter focuses on combining several techniques to establish a new method of generating a summary of reviews in an efficient and effective way.

1.3 Different Level of Analysis

- **Document level**: In this approach, the whole document is considered to be a single entity and the analysis approach is applied to the whole document. At the document level, to obtain an overall opinion value for the complete document, the task of research at this level is to classify the opinion because the overall meaning, that is the conclusion of the sentiment that has been expressed, is it a positive, negative, or neutral sentiment. For instance, if we have a product review, the system determines the opinion normally whether or not the review in general was evaluated to express positive, negative, or neutral sentiment. This type of research is called document-level sentiment classification.

  Example

  > I bought an i phone Some days past. It's such an awesome phone, though a little large. The touch screen is cool. The voice quality is evident too. I simply love it

  Is the review classification positive or negative? Document level classification works best once the document is written by a single person and expresses an opinion/sentiment on one entity.
- **Sentence level:** In the sentence level approaches, every sentence is considered as an entity and analysis is applied to each individual sentence and then the results are summarized to provide the overall result of the document. This is known as clause level analysis [12]. In this type of level of analysis, the analysis is concerned with the sentences and determining whether or not the polarity of the sentence is positive, negative or neutral. Neutral means no opinion in the sentence. The sentence level of the study is closely associated with subjectivity classification, which distinguishes sentences that express subjective views and objective analysis of specific factual data from the sentences. In every review, it is decided whether or not every opinion sentence is positive or negative.

  Example

  > iPhone sales are doing well in this bad economy.

  Sentiment classification at both the document and the sentence level is useful, but they do not find what people like or dislike, nor do they identify opinion targets.
- **Entity and aspect level:** As mentioned above, the usage of document level and sentence level analysis is not precisely determined and does not explore precisely what individuals like and dislike of a product's options. The aspect (feature) performs an ideal analysis in some cases. The primarily feature-based level or aspect-based level as it has been referred to recently, looks directly at the opinion itself rather than at an aspect of the target that is mentioned in each sentence, because opinion holders comment on different target options or aspects and it is not honest to evaluate an opinion overall.

Example

> Samsung mobile phone camera is fantastic, and it's a really nice design, the memory isn't enough for many applications.

Referring to the example, it cannot be completely decided whether the voice of the customer is positive or negative. The first two sentences tend to give positive comments about the camera and style, whereas the last one is a negative opinion regarding memory. In several remarks on a particular target, opinions are represented by target aspects or entities. Thus, the objective of the aspect level is to obtain a response on every entity and the aspects that are mentioned throughout the sentence. This is challenging. However, a regular opinion only expresses a sentiment on a particular entity or an aspect of the entity. e.g. "the mobile battery is bad," which expresses a negative sentiment about the aspect battery of a mobile.

1.4   Text Summarization Approach

A text summarisation approach is often classified into two categories: extraction and abstraction. Extraction outline could be a choice of sentences or phrases from the initial text with the very best score and it is placed alongside a replacement shorter text while not ever changing the supply text. Abstraction outline methodology uses linguistic strategies to look at and interpret the text [10]. Two alternative classes are often classified as fusion and compression. Fusion combines extracted elements coherently and compression is aimed at throwing out unimportant sections of the text. The summarization method has three phases: analyzing the supply text, determining its salient points, synthesizing an applicable output. An outline is often indicative, informative, or critical:

- Indicative summaries follow the classical information retrieval approach: they supply enough content to alert users to relevant sources that users then browse in additional depth.
- Informative summaries act as substitutes for the supply, in the main by collecting relevant or novel factual information during a summary structure.
- Vital summaries generated in a output provides informative gist, opinion Statements on content. They add worth with an up-to-date transferral experience that is not on the market from the supply alone.

1.5   Porter Stemming Algorithm

Natural language texts generally contain many variants of a basic word. Morphological variants (e.g. process, computer, computers, computing etc.) of the area unit are typically the most common, with alternative sources and valid different spellings, misspellings, and variants arising from written text and abbreviation. The advantage of this algorithm is looking out, most clearly, however, not solely in terms of recall, would be expected to extend if it were feasible to mix (i.e., to bring together) the variants of a given word so that they might all be retrieved in response to a question that such as simply one variant [18]. In English, and in many connected languages, morphological variation takes place at the right-hand finish of a word-form and this has spurred the utilization of user directed right-hand truncation for on-line information retrieval. This is often a very easy approach to conflation; however, one that needs a good level of expertise, as two major sorts of error area unit are possible. Over-truncation happens once too short a stem remains after truncation and leads to entirely unrelated words being conflated to an equivalent root, such as "medical" and "media" each being retrieved by the foundation "MED*". Under-truncation, conversely, arises if too short a string is removed and leads to connected words being delineated by totally different strings, such as "bibliographically" being truncated to "bibliographic", instead of to the shorter root "bibliograph*". A totally machine-controlled different truncation is provided by a stemming. This reduces all words with an equivalent root to one type, the stem, by husking the foundation of its derivation and inflectional affixes; in most cases, solely suffixes that are additional to the right-hand finish of the foundation area unit are removed.

1.6 Areas of Applications of Opinion Mining and Sentiment Analysis

1. **Purchasing product or service:** Taking the right decision while opting for a product or service is much simpler nowadays. People can easily compare others' views in terms of opinion and experience with regard to a product/service and look for a competing brand. People nowadays tend to go with their own views instead of relying on a consultant. Sentiment analysis and opinion mining play major roles in extracting opinions from a huge collection of unstructured content on the Internet, analyzing it and then putting it in a highly structured and understandable format [6].
2. **Decision making:** People's opinion and experiences are important elements in the decision-making process. With the help of opinion mining and sentiment analysis, peoples' views can be efficiently used for decision-making.
3. **Opinion spam detection:** Since the internet has been available to all, anyone can put anything on the internet. This has increased the possibility of spam content on the web. People may write spam content to mislead other people. Opinion mining and sentiment analysis can classify the internet content into spam content and non-spam content.

4. **Recommendation systems:** By classifying people's opinions into positive and negative, the system can say which one should be recommended and which one should not be recommended
5. **Marketing research:** The result of sentiment analysis techniques can be utilized in marketing research [18]. By sentiment analysis techniques, the recent trend of consumers with regard to some product or service can be analyzed. Similarly, the recent attitude of the general public towards some new government policy can also be easily analyzed. All these results can contribute to collective intelligent research.
6. **Quality improvement in a product or service:** By opinion mining and sentiment analysis, the manufacturers can collect critical opinions in addition to the favourable opinions about their product or service and they can thereby improve the quality of their product or service [5].

1.7   Research Problem Formulation

Sentiment analysis techniques are becoming popular in everyday life applications appearing on the web. Sentiment analysis is a sub-branch of text mining and opinion mining. It can be considered a hybrid approach. The input text is analyzed for polar analysis to mark it either positive or negative. The public spread their views over the social networks. The sentiment analytically approached can be knowledge-driven or unsupervised and supervised. The supervised sentiment analytical models have been found to be more effective than the other learning models. The linguistic efficiency of the dictionary words decides the weight of the message, which can be determined by online linguistic dictionaries equipped with weight information. The WordNet and SenticNet are the most famous and popular among other alternatives. This chapter focuses on the sentiment analysis and concludes the shortcomings of the existing models to improve the design of the sentiment analysis models. The proposed model also utilizes the Porter stemming, which is the method of dictionary reduction by using the Porter stemming module. The dictionary size reduction method is incorporated to produce quick results of the sentiment analysis.

1.8   Objectives of The Study

The objective of this research work is to provide customers with an optimized summary. This work mainly focuses on a number of reviews available on e-commerce sites. Every client wants to read views about a specific product before placing an order, but it is difficult to go through all reviews and focus on each one individually. Thus, this system provides a graphical summary of all the reviews related to the chosen product. A summary is a collection of all the features of a product and illustrate what is positive and negative about each feature. Thus, it helps the client to make a decision before buying a product and easily make a comparison of two products. Although much work had been done in this field, this system provides better results with certain modifications. The main objectives are as follows:

1. Porter stemming is designed to overcome the shortcomings of the existing techniques. It is integrated into the sentiment analysis module to build the complete sentiment analysis model.
2. The social dataset analytical module is implemented for the sentiment analysis.
3. An optimized summary text is provided in addition to an image of customer reviews that gives each feature summary of a product, not as a whole.

To achieve the objective of the method, this approach requires customers to select what kind of mobile phone they want to purchase and choose one other to compare the two. The proposed method provides text and an image summary of both mobiles that show the various features of each product from the reviews written by other people.

2   Related Work

This section describes the existing approaches to opinion mining and text summarization techniques. Different authors work in various fields of sentiment analysis, some of which are described below.

Cruz et al. [8] defined a group of domain-specific resources to extract the opinion words. Domain-specific resources consist of a group of options such as feature taxonomy, feature cues and dependency patterns. They utilized mostly lexicon-based approaches such as the WordNet, PMI and SentiWordNet-based classifiers for sentiment classification. The lexicon used the random walk algorithmic program. Pang et al. [17] proposed a model by using unigram, bigram, and adjectives to generate a feature vector for various machine learning algorithms on a dataset of movie reviews. Liu et al. [14] automatically created the fuzzy domain sentiment ontology tree (FDSOT), primarily based on the product options and sentiment words. The FDSOT was used more for features primarily based on sentiment classifications. Dang et al. [9] classified sentiments with the support vector machine (SVM) classifier by using different ways of selecting features. Xia et al. [20] considered electronic reviews using the SVM classifier and proposed an approach to the problem of noisy features due to ensemble features. Matsumoto et al. [15] implemented a sentiment model with the SVM classifier by using various features such as unigrams, word substitutes and a dependency subtree with a frequent subtree mining algorithm. Abbasi et al. [1] proposed a combinational approach of a genetic algorithm (GA)and information gain (IG). This model was implemented on movie reviews using SVM classification. Tu et al. [19] developed an approach to extracting very complex features using a dependency feature with unigram feature extraction. They evaluate the performance for the MaxEnt classifier. Agarwal et al. [3] developed a feature extractor model by using dependency relation with words present in the text and also creating a new technique for the removal of redundant information. They evaluate their system by using movie and electronics reviews. Mullen et al. [16] proposed a hybrid method of using Turney and Osgood values. They worked on analysis in movie reviews with hybrid SVM classification. Bespalov et al. [4] proposed efficient embedding for modelling higher-order (n-gram) phrases that projects the n-grams to low-dimensional latent semantic space, where a classification function can be defined. Che et al. [7] have worked on the sentence compression for aspect-based sentiment analysis. The authors have addressed the matter of sentiment analysis by proposing a framework of adding a sentiment sentence compression step before playing the aspect-based sentiment analysis. Flores et al. [11] described stemming algorithms in four languages, in terms of accuracy and in terms of their aid to data retrieval. The aim is to assess whether or not the foremost correct stemmers are also those that bring the most important gain in data retrieval. Experiments in English, French, Portuguese and Spanish show that this is often not invariably the case, as stemmers with higher error rates yield higher retrieval quality.

3 Flow of the Proposed Approach

The proposed model has been designed in various components. The proposed model contains various modules such as tokenization, stemming, sentiment estimator/calculator, stop word filter etc. Each component has its own design and way of working. All of the components have created the final model of the proposed work based on the stemming-based mobile phone review sentiment analysis. The flow of the proposed approach is shown in Fig. 1 The following main steps show the proposed methodology.

1. **Post/thread acquisition:** The first step is to obtain the data from the database to read all the reviews and all are analyzed step by step.
2. **Tokenization:** Tokenization is the process of extracting the token from the message data. The tokenization is based upon the n-gram analysis of the emotion data. The n-gram analysis defines the length of the keywords being extracted under the process of tokenization. In this part, the code reads recognized words from user comments based on the word-list prepared referring to commonly used words. This word-list is prepared by matching the words in the user comments along with a suitable word-list already saved as a text file. The text file is then loaded into the memory and the tokenization process forwarded for supplementary computations. In tokenization, all the words from user comments are extracted and filtered based on a list of frequently used words without emotions. Matched words in the user comments are deleted and the filtered list becomes ready to use. These common words are not given in the word weight file containing the rank/weight of each word being used in the common English language, which contains a neutral, positive or negative emotion.

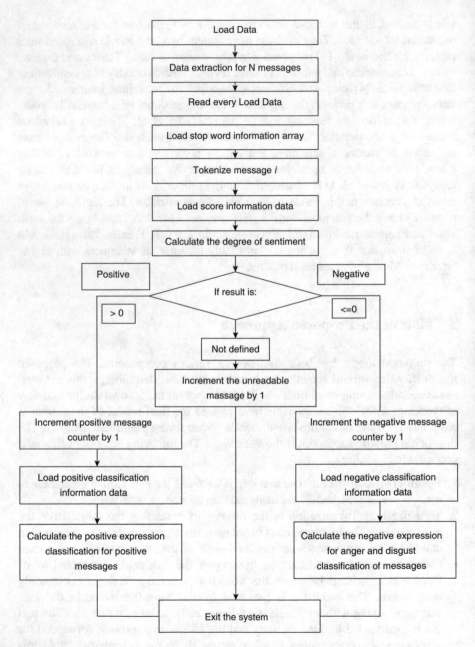

Fig. 1 Flow chart of the proposed approach

3. **Stemming:** Stemming is the process of finding the root words from the non-root words. All of the tokenized words that carry any grammatical or non-grammatical effects have been programmed to undergo the stemming procedure, which is programmed as Porter stemming. Porter stemming analyzes the word for the any grammatical or non-grammatical suffix, and removes the suffix from the word to convert it to the root word. The root word is the stem word from all forms of the word, which can be also considered to be the word with a neutralized effect. The stemming process returns the stem word, which carries no emotional weightage. For example, "love" expresses the normal expression whereas the word "loving" carries the higher order of emphasis or weightage and spreads a higher order emotion of "love". This effect of emotional emphasis is neutralized using Porter stemming in the proposed model.

4. **Sentiment Analyzer:** This module is the emotion mining module in the proposed model, which analyzes the emotion or sentiment of the given message after tokenization and the stemming procedure. The emotion is returned in the form of the numerical weightage of each word followed by the summation of the emotional weightage of each word in the whole message to know the overall emotion of the word. Then, the word is tagged with a positive or negative based on the emotional weightage.

5. **Polarization (positive or negative rating module):** The customer reviews are defined based on their polarity. It can be either positive or negative depending on the words appearing in the reviews. In this step, all the words from a tokenization step are compared with an existing list of words according to which polarity is assigned to each review. Weightage is assigned to each word between $+5$ and -5.

6. **Negative emotion analysis:** All the user reviews that are assigned a negative polarity are further analyzed to represent the extent of negativity in a comment. Messages are further divided into messages showing anger and messages showing disgust. The comments are marked based on higher weight. For example, if a public review of the specific product is found to be highly positive, then the text summarization module shows the significant emphasis on the emotion analysis. All these comments are further analyzed in depth for anger and disgust score files.

4 Experimental Results

4.1 Dataset Description

Reviews are collected from an online source of e-commerce sites such as Flipkart, Snapdeal and bestbuy. The reviews are collected for six products such as smart phones. The proposed model shows the polarity comparison between any two products chosen. This helps the customer to easily select a particular product to buy by reading a text summary of reviews posted by other people who have already used that product. This model also provides an image summary of reviews because nobody has time to read all the reviews. The following are the mobile phones whose data were collected:

1. NOKIA LUMIA-920
2. NOKIA LUMIA-1020
3. SAMSUMG EDGE-S6
4. SAMSUNG-A5
5. SONY-XPERIA-Z1
6. SONY-XPERIA M4

4.2 Random Forest Classifier

Random forests is a classifier that is based on a technique of classification that grows many classification trees to overcome the instability of previous decision trees. To classify a new object from an input vector, this input can be run through every tree in the forest. Every tree gives a classification and votes for the most popular class. The forest chooses to classify the case according to the label with the most trees in the forest. This classifier is non-parametric, so there is no need to think about the linearity of the input dataset. If parameters are present, they can be easily entered, thus eliminating the need to prune the trees. In the proposed model, the random forest works to classify the review into various parameters.

4.3 Algorithm

The product review analysis using opinion and text mining approaches can play a vital role in the e-commerce industry by producing detailed product reports using the in-depth analysis of the product reviews. The classification and analysis of the product reviews reveal the public opinion about the various physical or logical properties of the products undergoing the analytical study under the proposed model. The mobile phone data have been evaluated under the product review mining project in the proposed model, which includes popular handsets from Samsung, Sony and Nokia. The following algorithm shows the overall working of the proposed model over the given data:

Algorithm 1 Review analytical algorithm

1. Acquire the complete data from the input review corpus
2. Acquire the selective data by extracting the message or review data from the input product review corpus
3. Read the thread and load it into the runtime memory for further processing.

4. Estimate the size of the input corpus data after applying the pre-processing mechanisms.
5. Acquire the STOP WORD corpus containing the word list for early elimination.
6. Iterate the system for every message with the following sub-steps

- Obtain the messages based on unique id in the given corpus
- Apply the STOP WORD and eliminate the words matching the stop word list
- Apply the token extraction (also known as tokenization) process to the input data
- Stem the tokens to their root words using the Porter stemming algorithm
- Hand over the token data to the score module, which loads the dictionary set and computes the individual token score.
- Compute the overall score from the token score list
- Determine the depth of the sentiment in the given message from the degree of emotion in the given message text
- Define the polar tag over the message and add to the appropriate list
- Estimate the existence of the anger, disgust and the two emotions in the given text

The method of classification and categorization plays an important role in the generation of the product review reports based on the different properties and specific features of the products. The proposed system has been utilized with the automatic review classification and categorization algorithm, which has been elaborated in the following algorithm:

Algorithm 2 Automatic review classification algorithm

1. Acquire the classification weightage data along with the keyword list for the various representations of the various product categories Extract the individual weight information for each of the categories defined in the training data for review classification
2. Obtain the message data
3. Extract the keywords from the data according to the keywords representative of the input features
4. Find the weights for the keywords extracted from the message
5. Determine if the number of keywords is greater than 0

    a. Determine the category based on the existing weight codes
    b. Determine if the weight code of any of the review matches the category code
    c. Classify the message under the particular category

6. Aggregate the category information
7. Return the category information

5 Quantitative Evaluations

Sentiment analysis has been carried out based on the database consisting of reviews of customers to extract emotions automatically. Product feature classification is done after analysis of the messages. The emotions are analyzed based on words defined by polarity and weightage is assigned to each word. Messages are first divided into words, which are called tokens, and tokens are matched with a pre-programmed dictionary used in the proposed model. The quantitative evaluation represents the results based on the quantity of the entries in the given dataset. The proposed model has been evaluated for the variety of the parameters under this evaluation study. The total positive ratio, total positive and negative messages, messages representing disgust, anger and both anger and disgust-based options have been selected under this study. The proposed model has been designed to compare the two datasets obtained from the various sources in terms of their customer satisfaction based on the aforementioned parameters. The results obtained for the Samsung S6 edge and Sony Xperia Z1 models have been represented in Table 1. The total positivity ratio obtained for the Samsung model was 0.641 (64.1%) and 0.5493 (54.93%) for the Sony model. The total numbers of messages, 78 and 71 respectively, have been evaluated for the Samsung and Sony model. The total numbers of messages of 50 and 30 respectively have been found to be positive for the Samsung and Sony models, whereas there were 28 and 33 negative messages respectively for the two models. Seventeen anger and 0 disgust messages have been found for the Samsung model, whereas 1 message containing disgust and 14 containing anger messages were found for the Sony model. Only one message with both anger and disgust has been found in this study for the Sony model compared with zero messages for the Samsung model. The mobile phone analytical engine is based upon the one after one processing of both of the databases. The individual databases are analyzed and categorized with regard to the different features, including battery, screen quality, sound, network etc. The proposed model offers the inter-entity comparison of both of the input databases to be performed after analyzing the individual features of both of the input mobile phone handsets. The feature-based analysis empowers the

Table 1 The proposed model showing results using sentiment analysis

Index name	Samsung-S6-Egde	Sony-Xperia-Z1
Total positivity ratio	0.641	0.5493
Total messages	78	71
Total positive messages	50	39
Total negative messages	28	33
Disgust	0	1
Anger	17	14
Both	0	1
Unreadable messages	0	0

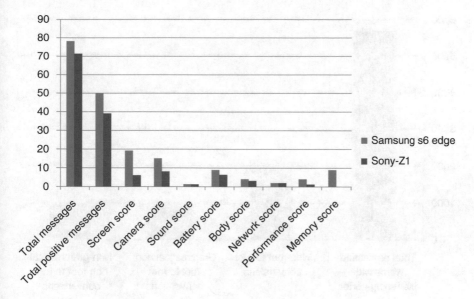

Fig. 2 The overall feature-based comparison

analytical system to deeply analyze the emotion of the users across the input dataset as shown in Fig. 2. This shows only positive messages regarding every feature of each product.

5.1 Porter Stemming Algorithm

The proposed model has been deeply analyzed for the performance of the Porter stemming model. The Porter stemming model has been designed to work in dictionary-based object matching before converting the words to the root words. The proposed model Porter stemming has been analyzed for the various types of encounters performed by the porter model. The major functions of Porter stemming such as the total number of processed words with Porter stemming, non-root-to-root conversions, grammatical non-root-to-root conversions and non-grammatical non-root-to-root conversion have been analyzed in Fig. 2. The total number of analyzed words are displayed with the range of nearly 5600–5700 words in the above figure, out of which approximately 1900 words are converted to the root words. Nearly 900 words were found to be grammatical and 1000 words were found to be non-grammatical according to the comparison of the two mobiles, as shown in Fig. 3.

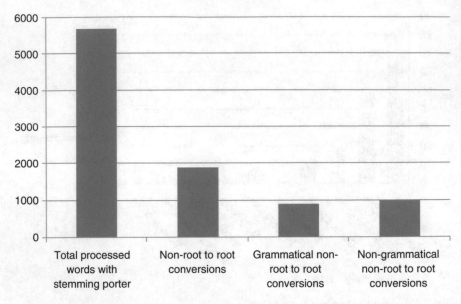

Fig. 3 Result analysis report obtained from Porter stemming

Table 2 Sentence compression rule-based accuracy analysis

Compression rule type	Accuracy value
Without compression	94.14
Manual compression	93.75
Automatic compression module	96.23
Average accuracy	94.71

6 Results Analysis

In this research, analysis is carried out based on Smartphone reviews collected from e-commerce sites. The proposed sentiment analytical model can classify the reviews into multiple parameters such as positive, negative, anger, and disgust. A dictionary of words is utilized to classify them. The modified random forest classifier is used to classify the polarity of the reviews. The proposed model has been measured for the different types of sentence compression methods over the product review data. The proposed model has been calculated with nearly 94% accuracy using the without compression module, whereas the manual compression model has been recorded with 93.75% accuracy, as shown in Table 2. The proposed model has accuracy of 96.23% while using the automatic compression module. Also, the proposed model has been analyzed for the average accuracy of the sentiment analysis model for all three types of sentence compression models. The proposed model has been recorded with 94.71 accuracy for the electronics category, where the mobile phone reviews have been analyzed under the proposed sentiment analysis algorithms. The proposed model has performed better than the existing models of Basant Agarwal's

[3] scheme (88.9%) and Dang's scheme [9] (83.75%) based on the overall accuracy, which has been recorded as the best value from the performance evaluation rounds. Our proposed model has even performed better overall when compared with all other models, irrespective of the review category, as shown in Table 3. The proposed model has been evaluated as being the improved and most accurate model compared with all the other models evaluated under the comparative analysis of the proposed model. In Table 4, the sentiment analytical model has been analyzed for the different

Table 3 Comparative analysis among the existing and proposed methods

Paper	Machine learning algorithm	Dataset	Best accuracy
Cruz et al. [8]	RF	Hotel reviews	93.45
Pang et al. [17]	SVM	Movie reviews	87.1
Liu et al. [14]	FDSOT	Laptop reviews	70.0
Dang et al. [9]	SVM	Movie reviews, DVD, electronics	78.85, 80.75, 83.75
Xia et al. [20]	Ensemble classifier	Movie reviews, books, DVD, electronics	88.0, 83.0, 83.8, 86.0
Matsumoto et al. [15]	SVM	Movie review	92.9
Abbasi et al. [2]	SVM	Movie reviews	91.7
Tu et al. [19]	MaxEnt	Movie reviews	91.6
Basant Agarwal et al. [3]	SVM	Movie reviews, books, DVD, electronics	90.1, 88.5, 89.2, 88.9
Mullen et al. [16]	Hybrid SVM	Movie reviews	86
Abbasi et al. [1]	SVM	Movie reviews	87.9
Xia et al. [21]	SVM	Movie reviews	88.6
Our approach	Modified random forest	Electronics (smartphone reviews)	94.71

Table 4 Comparison of the proposed and existing models with regard to various errors

Parameter type	Existing sentence compression and sentiment analysis model [7]		Proposed sentence compression and sentiment analysis model	
	Correct	Incorrect	Correct	Incorrect
Aspect polarity collection	390	118	425	93
Error produced by the algorithm	15.6%	17.8%	10.2%	14.3%
Errors produced in syntactic context	12.3%	27.1%	8.9%	25.0%
Error produced with wrong compression	0%	20.3%	0%	17.5%
Total accuracy analysis	72.10%	34.80%	82.00%	28.60%

types of models for sentence compression and sentiment analysis. The sentence compression model returns the keywords extracted from the input text data, which are further used for the assessment of the polarization of the text messages. The proposed model has been analyzed for the correct and incorrect results produced by the sentence compression methods in the existing and proposed models. The proposed model has outperformed the existing model, with 72.10% accuracy and nearly 82% accuracy for sentence compression, which plays a vital role in the sentiment analysis model. The proposed model has produced 28.60% incorrect sentence compression results, which is significantly better than the existing model, with 34.80% overall accuracy for the existing aspect-based approach to sentence compression.

7 Conclusion

The proposed model has been designed for adaptive keyword extraction for the new and improved sentiment analysis model. The proposed model has been developed by utilizing the modified random forest model, which correlates the keywords and produces accurate sentiment analysis results. The proposed model is better than the existing model, with 94.71 % accuracy, which is clearly higher than that of the existing model of 88.9% and 83.75% for the review analysis over the product reviews related to the electronic products. Also, the proposed model has outperformed all of the existing models, which have been evaluated over the other types of product reviews such as movie reviews, CD/DVD reviews and book reviews based on the overall accuracy using the proposed sentiment analysis model. Furthermore, the proposed model is more efficient than the existing model with regard to sentence compression, where it has not violated the null percentage readings and returned the null or zero value of a similar property, such as no error has been produced under the error produced with the wrong compression category as shown in table 4. In addition, the proposed model has outperformed the existing models based on the syntactic and algorithmic error categories, where nearly 10 and 14% accuracy has been recorded for correct and incorrect readings respectively, compared with the 15.6 and 17.8% accuracy of the existing model for correct and incorrect readings under the algorithmic error category. Finally, nearly 9 and 25% accuracy has been recorded for correct and incorrect readings respectively, compared with 12 and 27% accuracy of the existing model regarding correct and incorrect readings under the syntactic error category.

References

1. Abbasi, A.: Intelligent feature selection for opinion classification. Technology **54**(14), 1269–1277 (2003)

2. Abbasi, A., Chen, H., Salem, A.: Sentiment analysis in multiple languages: feature selection for opinion classification in web forums. ACM Trans. Inf. Syst. (TOIS) **26**(3), 12 (2008)
3. Agarwal, B., Poria, S., Mittal, N., Gelbukh, A., Hussain, A.: Concept-level sentiment analysis with dependency-based semantic parsing: a novel approach. Cogn. Comput. **7**(4), 487–499 (2015)
4. Bespalov, D., Bai, B., Qi, Y., Shokoufandeh, A.: Sentiment classification based on supervised latent n-gram analysis. In: Proceedings of the 20th ACM International Conference on Information and Knowledge Management, pp. 375–382. ACM, New York (2011)
5. Brisson, L., Torrel, J.-C.: Opinion mining on experience feedback: a case study on smartphones reviews. In: 2015 IEEE 9th International Conference on Research Challenges in Information Science (RCIS), pp. 187–192. IEEE, Piscataway (2015)
6. Cambria, E., Olsher, D., Rajagopal, D.: Opinion mining and sentiment analysis -challenges and applications. In: International Journal of Application or Innovation In Engineering and Management (IJAIEM), pp. 401–403 (2014)
7. Che, W., Zhao, Y., Guo, H., Su, Z., Liu, T.: Sentence compression for aspect-based sentiment analysis. IEEE/ACM Trans. Audio Speech Lang. Process. **23**(12), 2111–2124 (2015)
8. Cruz, F.L., Troyano, J.A., Enríquez, F., Ortega, F.J., Vallejo, C.G.: 'Long autonomy or long delay?' the importance of domain in opinion mining. Expert Syst. Appl. **40**(8), 3174–3184 (2013)
9. Dang, Y., Zhang, Y., Chen, H.: A lexicon-enhanced method for sentiment classification: an experiment on online product reviews. IEEE Intell. Syst. **25**(4), 46–53 (2010)
10. Ding, X., Liu, B., Yu, P.S.: A holistic lexicon-based approach to opinion mining. In: Proceedings of the 2008 International Conference on Web Search and Data Mining, pp. 231–240. ACM, New York (2008)
11. Flores, F.N., Moreira, V.P.: Assessing the impact of stemming accuracy on information retrieval–a multilingual perspective. Inf. Process. Manag. **52**, 840–854 (2016)
12. Garcia, D., Schweitzer, F.: Emotions in product reviews–empirics and models. In: 2011 IEEE 3rd International Conference on Privacy, Security, Risk and Trust (PASSAT) and 2011 IEEE 3rd International Conference on Social Computing (SocialCom), pp. 483–488. IEEE, Piscataway (2011)
13. Kherwa, P., Sachdeva, A., Mahajan, D., Pande, N., Singh, P.K.: An approach towards comprehensive sentimental data analysis and opinion mining. In: Advance Computing Conference (IACC), 2014 IEEE International, pp. 606–612. IEEE, Piscataway (2014)
14. Liu, H., He, J., Wang, T., Song, W., Du, X.: Combining user preferences and user opinions for accurate recommendation. Electron. Commer. Res. Appl. **12**(1), 14–23 (2013)
15. Matsumoto, S., Takamura, H., Okumura, M.: Sentiment classification using word subsequences and dependency sub-trees. In: Pacific-Asia Conference on Knowledge Discovery and Data Mining, pp. 301–311. Springer, Berlin/Heidelberg (2005)
16. Mullen, T., Collier, N.: Sentiment analysis using support vector machines with diverse information sources. In: EMNLP, vol. 4, pp. 412–418 (2004)
17. Pang, B., Lee, L.: A sentimental education: Sentiment analysis using subjectivity summarization based on minimum cuts. In: Proceedings of the 42nd Annual Meeting on Association for Computational Linguistics. Association for Computational Linguistics, p. 271(2004)
18. Porter, M.F.: An algorithm for suffix stripping. Program **14**(3), 130–137 (1980)
19. Tu, Z., Jiang, W., Liu, Q., Lin, S.: Dependency forest for sentiment analysis. In: Natural Language Processing and Chinese Computing, pp. 69–77. Springer, Berlin/Heidelberg (2012)
20. Xia, R., Zong, C.: Exploring the use of word relation features for sentiment classification. In: Proceedings of the 23rd International Conference on Computational Linguistics: Posters, pp. 1336–1344. Association for Computational Linguistics (2010)
21. Xia, R., Zong, C., Li, S.: Ensemble of feature sets and classification algorithms for sentiment classification. Inf. Sci. **181**(6), 1138–1152 (2011)

Trends and Pattern Analysis in Social Networks

Meenu Chopra, Mamta Madan, Meenu Dave, and Cosmena Mahapatra

Abstract The focus in India has now changed to providing world class higher education to all students aiming to compete with the world. Keeping this in mind, higher education institutions (HEIs) must figure out a way to make the change in accordance with technological advancements in social networking to motivate students and encourage an intuitive learning environment in their campus. Using social networking in *higher education (HE)* is being creative, productive, cost–effective, and is exceptionally critical owing to the complex nature of serving the population of the *online generation (Og)*. Utilizing the *social networking environment* for learning and teaching at HEIs could be a financially effective and productive way of speaking to and connecting with higher education online members, which include students, faculties, administrators, staff, management, etc. A few illustrations of the HEI becoming a "Social Institution" are *strengthening the HEI's "brand" or reputation, managing and building online communities (staff, students, parents/guardians, graduated class/alumni), and streamlining processes for better productivity at less expense*. This chapter focuses on effective implementation of data analytics techniques on social media datasets for helping *Indian HEIs* to compete effectively in the global market.

Keywords Higher education (HE) • Higher education institutions (HEIs) • Higher education professionals (HEPs) • Online generation (Og) • Online social networking (OSN) • Social networking sites (SNS)

M. Chopra (✉) • M. Madan • C. Mahapatra
Vivekananda Institute of Professional Studies, New Delhi, India
e-mail: meenu.mehta.20@gmail.com; mamta.vips@gmail.com; cosmenavips@gmail.com

M. Dave
Jagannath University, Jaipur, Rajasthan, India
e-mail: meenu.s.dave@gmail.com

1 Introduction

The role of *Online social networking* in academia is advancing rapidly, in step with increasing adoption across the wider society as a whole. Today, we find that students, parents, alumni, higher education professionals (HEPs) such as professors, teachers, and staff are embracing online networking, along with the almost ubiquitous use by students. While academicians discover the pedagogical uses of online social networking, higher educational institutions around the world are looking to find new or innovative ways to put to use online networking outside the classroom. Online social networking can deliver significant advantages, and even help to streamline the primary concern, especially in competitive private education and other affected organizations of higher education.

In today's world, the *higher education institutions (HEIs)* are continually adjusting to current patterns, keeping in mind the goal of improving students' learning environments. As the Indian educational culture accommodates a continually changing technological era, HEIs must figure out how to adapt and become the best academic institutions to meet the specific end goal of encouraging an intelligent learning environment. For a huge number of individuals, the Internet is a social hibernation space that interfaces them with their peers, colleagues, friends, relatives, and even aggregate outsiders. The new era of the *WWW*,[1] as it has been named, supports social communication and makes it simple for HEIs to trade data and to share their campus life and academic interests in some structure (e.g., sharing tutorial recordings on YouTube,[2] college fest photographs on Flickr,[3] contemplations in Blogger (service),[4] and aptitude in Wikipedia.[5] Therefore, the *online social networking (OSN)* of HEPs is more than just innovation.

The HEIs are dealing with pressure to retain more students, and to enhance the learning environment for the students. Innovation in technology has played a vital role in improving the academic culture by means of the virtual classrooms [19], collaborative advancement groups, and guaranteeing better correspondence with their students. Among all the *social networking sites (SNS)*, *Facebook* (see footnote 1) is the largest and most populated social network in the world to date, which has given a platform for and permitted *higher education (HE)* students to

---

[1]The World Wide Web, or the Web, is an information space where documents and other web resources are identified by uniform resource locators (URLs), interlinked by hypertext links, and can be accessed via the Internet.

[2]https://www.youtube.com. It provides a forum for people to connect, inform, and inspire others across the globe and acting as a distribution platform for original content creators and advertisers large and small.

[3]https://www.flickr.com, almost certainly the best online photo management and sharing application in the world.

[4]www.blogger.com, a blog-publishing service that allows multi-user blogs with time-stamped entries.

[5]https://en.Wikipedia.org/wiki/Wikipedia. Wikipedia is the largest and most popular general reference work on the Internet.

become a solid voice in their community. The heads of *HEIs* must be sensitive to their students and other*HEPs)* such as faculties, management, administrative staff, etc., needs and utilizing online social networking at HEIs could be a financially savvy and proficient way of speaking with and connecting with HE students and professionals. While doing this type of research, we believe that analyzing the patterns of students' behavior in the higher education-related social networks, we can produce information using hybrid teaching and learning environments in the campus. The intelligence of social networks can be further used for collaborative bonding between higher education and SNS usage. What we analyzed by detecting these patterns was that these SNSs are *personal learning networks, open educational resources, and massive open web-enabled online courses.* Aspectual SNS growth in the education industry is convenient and practical. According to the literature survey, most of the researchers believed that students access their Facebook (see footnote 1) account mostly to look for casual posts, messages or comments on current social affairs from their connected members or friends, but to have an optimistic view, we believe that they are also using Facebook (see footnote 1) to obtain updates on academic concerns and issues.

2 Literature Survey

Qualitative research focuses on the esteem in which undergraduate students hold social scholastic experiences, for which hybrid social and scholastic components are within the similar set of interactions [2–4]. Such findings are consistent with earlier studies on community college students [7, 12, 13, 28] and [27], which found that online communities of college students tend to understand and take advantage of integrative experiences that have a social component yet revolve around their scholastic interests. In addition, reconciliation or integration is less frequently portrayed by participation in social organizations than through information networks that often originate in and develop from classroom-based structures [13]. Integrating a physical classroom environment, especially through learning communities, has also been shown to increase overall social integration as well [27, 28]. Clearly, web-based social networking has the potential to provide a platform for such system- and relationship-building opportunities. It can be utilized as an instrument for augmenting the engagement of classroom communities into the online medium.

For instance, incorporating the formation of classroom groups on Facebook (see footnote 1), facilitating question and answer sessions by means of online networking, and even watching class-delivered lectures and supplemental tutorials on YouTube, utilizing the comments section to generate more lively class discussion. The studies that straightforwardly investigated the potential use of the platform are even more applicable to college students. Among those that do research in this area, the assessment of the impact of online social networking on particular student accomplishments and achievement results is rare. There are many case studies which have concentrated on the connections between Facebook (see footnote 1) [8–10] and

Twitter[6] [11] usage and scholastic performance or on expanding student engagement and inclusion by using social media platforms [5, 6]. The scarcity of such result-oriented studies puts constraints on the ability of research around this topic to offer solutions for higher education institution policy or interventions that address the neglected needs of our most vulnerable student populations.

The *online social networking* platforms, for example, *Facebook* (see footnote 1), have appeared to be popular as a Web 2.0 tool amongst undergraduates by encouraging them to share about college events and occasions, keeping these users well-educated or informed. There are numerous institutions that no longer issue personal emails to the student, but have a wired link with their students through the institution fan page on *a social networking site, i.e., Facebook,*[7] the official page for the Vivekananda Institute of Professional Studies (VIPS). The online social networking permits students to share, impart, and exchange thoughts on the Web in addition to building logical associations within the communities in the social network. The *social networking* is a decent correspondence venue and a parkway where staff or heads, HEPs or alumni can make associations with students by taking a look into their lives as posted on Facebook (see footnote 1). Online social networking, in a specific term, accessing social network applications through the internet connection, is a new trend in almost every HEI today. Few researchers aroused the negative impact of student academic performance by using a social networking site during learning hours. We have seen in our literature survey that in a higher education environment the phenomenon of using social networking is massive. Although, the debate is still on about the negative impact of social networking on students' academic productivity, even during lectures, they are accessing these sites to disseminate information and communicate among themselves.

Nowadays, the faculties of HEIs have started using blogs as a pedagogical strategy. Recent research has investigated the use of blogs in scholarly disciplines including the sciences, language learning, teacher education, and business. The higher education professionals, especially faculty, have moreover made Facebook (see footnote 1) profiles to interface with their students in a more personable and casual space [26]. This has additionally prompted the employment of developing communities on Facebook (see footnote 1) as groups for course offerings that previously used web-based forums for discussion [27]. Although the popular perception may be that the use of online social networking for increasing pedagogical practices and experiences is restricted in respect of its utilization for different purposes (alumni relations, streamlining processes, enhancing reputations), no studies have actually documented the extent of each type of use in relation to each other.

A review of the existing literature on the use of social media in higher education provides a benchmark for the broad current uses, providing a descriptive overview of

---

[6]https://www.twitter.com, to give everyone the power to create and share ideas and information instantly, without barriers.

[7]https://www.Facebook.com/Vivekananda-Institute-Of-Professional-Studies

the phenomenon. Scholarly journals, research reports, blogs, individual university documents, higher education news, books, and online media were utilized as sources to conduct this review. Such a broad presentation of the facts, user practices and behaviors are relevant enough to support the fast-growing aspect of our communication patterns, and culture can perhaps begin to challenge the assumptions that some researchers and educational practitioners may have about how online social networking is used by HEIs and its stakeholders today.

We plan to utilize this review of the literature as a foundation that allows us to better capture the role and effect of online social networking among the HEPs, students, and HEIs. We also hope our undertaking of this task can help us to move toward the objective of HEI good practices that exist across peer HEIs considered viable and inventive. Next, we inspect our social networks for what past research tells us about the effect of social networking on higher education, especially the impact on students' improvement and character, learning, and other scholastic and social outcomes. Then, finally, we provide some suggestions for how social networking might be utilized to improve student achievement before addressing the challenges and potential perils of online social networking as it turns out to be increasingly pervasive among HEPs, students, and HEIs.

3   Social Networking in HEIs and Its Perspective

Currently, many HEIs are facing the task of administering and maintaining high-quality standards for higher education and enhancing the academic success of learners for their academic career. Although, there is an exponential increase in the enrolment numbers, and fewer dropout rates, academic accomplishments are still low. The cause of this is a major concern for HEIs that is still to be investigated. What is required is the framework or the model that regulates academic activity and defines the policies of the social media used in academia, especially, for those who need improvement and enhancement in their academic work. Our aim in this chapter is to combine the intelligence of online users' behavioral patterns of the HE-related social networks with the learning environment, to develop a framework or a pedagogy that uses social media for academic purposes. For this reason, we think that the HEIs should put time and research into gaining the benefit of the online networking environment. The HEIs should specifically define some academic policies on the use of technologies in the classroom for the online generation to cater for the needs of the students and learning environments of the HEIs.

We believe that the use of social media in academics will eventually lead to good performance among students and help HEPs, administrators, and management to run the institution in a more sophisticated way. However, before the implementation of social media in academia, it is essential for the protocol for HEIs to be formulated properly by laying down a specific mission and guidelines that require social media to be used as a tool for extra backing in academic and other HEI. The perspective of social media in education is to bring together and integrate all the members (students, administrators, HEPs, etc.,) of the HEI to increase productive activities beyond the classroom or campus. Also, the stakeholders of the institution must recognize the hurdles that any HEI faces. Research inferred that not only the students, but also the institutional academic tasks require a collaborative and coordinated effort from all its members to promote experimental application, career, and life success strategies. This coordinated effort comes into effect when HEIs are able to use social media in an effective way for the betterment of its learning environment. The online interpersonal interaction through social networking can be perceived as *web software applications or tools* for informal educational learning, which helps the students to learn within their associated communities in the social networks. Social networking helps and supports students to be self-coordinated and independently guide themselves toward specific learning ventures. A blend of both connections (community-oriented and free) might be perfect, while fusing online social media and networking systems into higher learning institutions or HEIs. This feeling of active and relevant engagement in social networking will propel their interest in academia and feeling of belonging to the institution.

Through social networking, students can experience a high level of engagement and interest, along these lines, giving them responsibility for their academic and professional career. Faculty members can utilize social networking to spur innovation in their respective teaching subjects. Through a literature review, we analyzed that most faculty members are now utilizing some sort of social networking, with more than 50% using online tools or web applications, for example, wikis, web journals, podcasts, and online tutorial videos, in the classroom. These social networking platforms are nowadays utilized for scholarly purposes and have demonstrated constructive and positive outcomes because of which students and faculty can interact outside the classroom. These online platforms, therefore, help in the learning process and in creating online communities. By integrating the physical (inside the classroom) and the virtual environment (outside the classroom), social networking has appeared to build shared and scholarly engagement, particularly for the freshers (newly admitted) students. Social networking and online community-based interactive platforms such as wikis, online bulletin boards, Facebook (see footnote 1), LinkedIn,[8] Twitter (see footnote 8), Skype,[9] YouTube (see footnote 5), web journals, are turning out to be increasingly predominant teaching tools for the classroom and HEIs. Facebook (see footnote 1) can be used in HE as a specialized interactive tool for the HEIs and stakeholders in addition to their guardians and peer group. We believe that a few factors can work for social networking in HE. They are: *consideration, cooperation, coordinated Effort, system mindfulness, and basic utilization.* As time goes by, social networking and its segments would expand these wings to reconcile to everybody's daily routine. Therefore, it must be fused into academia.

---

[8]https://www.LinkedIn.com, a business and employment-oriented social networking service that operates via websites.

[9]https://www.Skype.com is an application that provides video chat and voice call services.

Higher Education Professionals (HEP) need to understand the significance of interaction amongst students, guardians, friends, and other support administrative personnel in their first year of college and that is why a proactive policy on communication should be adopted in the campus environment. This fundamental online communication atmosphere builds a good working culture and also incorporates other factors such as *student retention, psychological advancement, active and relevant engagement, scholastic achievements, and interactive data mining patterns among institutions, students, and their families.* HEIs must recognize the significance of the online interaction among all those who involved, including social networking, to build working connections and associations.

The most important feature of social networking is openness. This openness is accompanied by risks such as individual well-being, security, and the loss of privacy. HEPs must see, talk about, and address how to handle the sensitive issues posed by the student. Maintaining the end goal, HEIs must monitor and keep a tab on the time needed to use online social networking. For this reason, a few technical or specialized solutions have been made to permit HEIs to react to remarks and posts by tagging certain catchwords or triggers that can then be sent to a maintenance team that takes corrective actions on a user's Facebook (see footnote 1) wall. The same web solution may address the issues that many HEPs have with fast-paced student interactions or communications with social networking platforms and accordingly mitigate a few wavers.

4 Methodology

The above-noted fundamental knowledge about the use of online social networks (the different ways in which social media are being used in higher education). More specifically, by whom and how frequently are social media being utilized within HE and in the wider real-world? Are there variations in user behavior patterns, and what recommendations or suggestions are there for different stakeholders in HE? To answer these questions, we further explore the social media usage patterns of HE-related networks on Facebook (see footnote 1). Online social networking is a world-wide global phenomenon embraced by most of the students, HEIs, and their professions. Based on this phenomenon, we conducted research to explore the use of social networking in the HE environment, especially among the faculty members and students, and analyze the impact on teaching-learning activity. We had evaluated few performance indicators [16] that led to today's higher learning institutions improvising, and also investigated the different relationships [22] that exist on *social media networks (SMNs)*: Facebook (see footnote 1) by using mining techniques. We studied various interactions within the social networks that play a crucial role in students learning through sharing and collaboration in the online content with their peers. In fact, we tried to understand the projection of these social networks in terms of various mathematical models [18]. From the HE perspective, we first sought to understand the latest technologies [17] that can be used on campus to give HE a face-lift by going social. It was also shown in [21] that the semester *cumulative grade point average (CGPA)* predicts that the frequency of using Facebook (see footnote 1) by students in their academic actions is high and the total time spent on Facebook (see footnote 1) for casual use is negatively correlated (linked) with CGPA. The research was conducted in three private universities that are familiar with social networking activities. The research focused on the usage of four kinds of activities, such as connecting to Facebook (see footnote 1), micro-blogging, instant-messaging, and blogging. A total of 300 respondents filled in the online survey. The result shows that most respondents agree to free access to social networking during office hours, and about 60% of respondents use this access not only to entertain but also for information distribution and communication to support teaching activities. The usage varies from task assignment, announcement, class rescheduling negotiation, examination, etc., in which some application such as Facebook (see footnote 1), Twitter (see footnote 9), instant messenger, and blogging sites is used. The policy of social networking access that is most suitable for user behavior in each environment should be proposed.

We have carried out much detailed research related to *social network analysis (SNA)* in our previous work [1, 15, 18, 20], in which we study social network models, their properties with respect to synthetic networks, and analysis that can be done on these social networks, especially Facebook (see footnote 1). By using *network analysis software applications (NASA) tools [25]*, we studied social networks and discovered community statistical information about them. For example, the Wolfram Alpha[10] application gives us the frequency of postings by each user. The SNA performed using NASA tools [25] gives us a picture of the Facebook (see footnote 1) social graph, which depicts how much each member participated, user positioning in the network (central, peripheral or outer), number of posts, their replies, and part of the specification used by members in the network. In the following section, we throw some light on SNA and its application in HE. In conclusion, we present the possible outcomes predicted by SNA carried out on the SNS, i.e., Facebook (see footnote 1) using NASA.

The reason for this subjective study was to find the importance of online social networking and the effect it has on the HE social networks of HEIs. The research focused on the study of engagement and behavior patterns of students of the four social networks on Facebook (see footnote 1) associated with one of the up and coming, growing and leading HEIs with National Assessment and Accreditation Council status "A", Guru Gobind Singh Indraprastha University. The reason for selecting these networks is because of the network user's topographical proximity and similarities in their demographics. The research focused on the HEI's networks, which were open, and that they have a high population of students, HEPs, management, administrators, alumni, training and placement officers (TPOs), etc.

---

[10]https://www.wolframalpha.com/ introduces a fundamentally new way of obtaining knowledge and answers, not by searching the web, but by performing dynamic computations based on a vast collection of built-in data, algorithms, and methods.

The study analyzed the different observations and perceptions that the network users have with regard to online networking and its impact on their behavior and engagement.

To obtain the outcomes for our study, we have used various software tools or applications. For example, *Netvizz*,[11] NodeXL,[12] Gephi,[13] WolframAlpha (see footnote 10), LikeAnalyzer,[14] which we termed as *network analysis software applications (NASAs)*, on higher education networks to collect the online network graph samples and to obtain the possible outcomes of our research. The universe selected for study is Facebook (see footnote 1), the largest SNS. The target population for our study is the four selected Higher Educational Social Networks (including students, HEPs, faculty, alumni, TPOs, administrators, management) were treated as samples: *Vivekananda School of Information Technology (VSIT) alumni (79 nodes and 889 edges), VSIT (367 nodes and 6468 edges), Bharati Vidyapeeth Institute of Computer Applications and Management (BVICAM; (42 nodes and 189 edges) and Higher Council of Education (1899 nodes and 20,296 edges).* Every network represents the solicited picture of an arrangement of users and their communities. We tried to pick up information about each individual or community social media impact on their engagements and academic patterns. By using the culmination of final outcomes or result information, we analyze the importance of online networking for HE within HEIs.

Our research study had used data samples of the HE networks (mentioned above) from the SNS Facebook (see footnote 1), which was taken a universe of study and used *NASA* tools on the four higher education networks, which are closed private networks especially for the students and professionals. The strength of the user population in each network extended from a few hundred users to 16,000. The proportion of students inside every network ranges from 41% to 93%. Every network met the basic protocols and therefore can be viewed as an HEI. The members of the network have common interests, subject streams, hobbies, etc., and therefore, were chosen with the expectation of permitting them to share their knowledge and academic experience in the college by using online networking. All four selected data samples were collected for the study and NASA tools were applied for their examination.

Of the four networks, two belong to *the Department of Information Technology of Vivekananda Institute, i.e., VSIT (Vivekananda School of Information Technology),* one belongs to *BVICAM (Bharati Vidyapeeth Institute of Computer Applications and Management)* and the last belongs to *Higher Council of Education*; the first

---

[11] https://apps.facebook.com/netvizz/. Netvizz is a tool that extracts data from different sections of the Facebook platform, in particular groups and pages, for research purposes.

[12] https://nodexl.codeplex.com/. NodeXL Basic is a free and open-source network analysis and visualization software package for Microsoft Excel.

[13] https://gephi.org/. Gephi is an open-source network analysis and visualization software package written in Java on the NetBeans platform.

[14] https://www.likealyzer.com/. It helps users to measure and analyze the potential and success rate of their Facebook pages.

three networks were institute-oriented and the fourth was generic. The HEPs (or those other than students) who belong to these social networks have many years of experience, ranging from 11 to 27 years either in teaching or in administering the HEI. All the faculty members in the online social networks have either a Master's degree or a doctorate, within their scholarly field. On the whole, they (HEPs, administrators, TPOs, management, etc.) had foundations working with the administration, student affairs, financial guidance, admissions, enrolment, placement activities, recruiting, learning assistance, and advising centers.

We have found through our studies that each member of the four social networks utilized some type (in any form) of online networking for academic work purposes. Likewise, three out of the four members have shown the utilization of online networking or social media in their personal and academic work on a regular basis. We found hardly any members, only one or two in each network, who do not utilize Facebook (see footnote 1) for their individual needs. All four Facebook (see footnote 1) social networks were using online social networking for academic and for personal reasons. The HEPs participated in the network discussion, with a specific end goal of giving their point of view, which is important, comprehensible, and ready to make an impact. Few new data mining patterns were picked from these HE-related social networks. Each pattern was precisely noted before a conclusion was reached and the guaranteed consistency checked amongst all the shortlisted networks. A few research questions were intended to obtain informational patterns with regard to the users of these higher education networks. The research questionnaire consisted of a few parameters such as demographics, age, active engagement inquiries, along with student's participation and academic-related questions.

5 Data Collection Using NASA

The online social media provide the basis for the individual establishment of each network. To check student determination for the active participation in social media networks, the study applied SNA to Facebook (see footnote 1) higher education pages or groups. For that, we extracted the sample networks related to HE groups in addition to student groups of HEIs from SNS, i.e., Facebook (see footnote 1), by using one of the NASA tools, i.e., Netvizz (see footnote 11) or NodeXL (see footnote 12). The following were the phases of our previous work, which is related to this current study:

5.1 Discovering Social Media Network

By using NASA tools such as *Netvizz* (see footnote 11), *NodeXL* (see footnote 12), *Gephi* (see footnote 13), WolframAlpha (see footnote 10), and *LikeAnalyzer* (see footnote 14) the database of the social network graphs was created, which contained attributes related to members' information (such as gender, label, posts, demographic information, etc.), and edge information (such as weight, directed or undirected, source, target etc.), we can obtain social media networks.

5.2   Evaluating the Active and Relevant User Engagement

In a network, if the edge (between two nodes) carries more weight (a high level of interactions or communication between nodes is high) in both types of network, directed and undirected, it indicates a high level of teamwork, learning process, good performance related to any assignment that has been given, social support by peer members and other social services. The functioning of the group depends on communication or interaction processing, active members' participation, activity or assignment processing, etc. Similarly, interaction of communications processing further depends on the interpretation, assessment, explanation, etc. Activeness, effectiveness, and relevant participation of the users or students in the network can be calculated or evaluated by calculating network properties such as betweenness, centrality coefficient, etc., which has been done in our previous work. Our research work focuses on how much and for what purpose each member in the higher education group is using the SNS (Facebook, see footnote 1) and whether this SNS should be added in the future as a classroom support tool or to the pedagogy, as SNS exchanges, have theoretically proved to be both a valid form of highly interactive instructional communication and an opportunity for pedagogical mentoring.

6   Research Questions

To analyze the connections between higher education-related social networks, online social networking, and its associations with students in academic accomplishment at HEI, the following set of research questions were chosen:

Question 1   To what extent or frequency of interactions are students using Facebook (see footnote 1) for academic learning or within their educational studies?

Question 2   What are the possible user demographics (age, gender, and locale) and if there are any gender differences in the groups for educational purposes?

Question 3   What are the possible types of associations (the commenter, friendships, initiators, etc.) and interactions (comments, posts, and likes) that exist in Facebook (see footnote 1) used?

Question 4   How many students are active, passive, and outliers in the group?

Question 5   How to calculate and compare the centrality of the students in the group?

Question 6 How to find weak and strong weight edges, communities, and representatives that exist in the network?

Question 7 What are the most probable terms or phrases or topics used in the network and how does network topology define itself in terms of communities within the group or network?

Question 8 What are the various activities performed by students in the group and by what time?

Question 9 What is the average number of messages or comments, sent by the active (members) students in the group?

Question 10 What are the major activities carried out by the students in the group?

We will look for the answers for the above-mentioned questions in the forthcoming sections and based on the results, we predict behavioral patterns and try to give the social networks hybrid intelligence.

7 Data Analysis

Data mining (DM) on social networks produces information that can be utilized as a steady relative model to help the researcher to find new patterns. The data mining of each of the four networks was recorded and translated verbatim for the *NASA*. Once the information in the format of graph files (.GDF extension) was deciphered, a suitable algorithm was applied to discover the key patterns, along with the current topics of discussion in the online social networking environment. The information was retrieved into an Excel table in the *NodeXL* (see footnote 10) *tool* and recorded for each network. After obtaining social network graphs, *SNA* was carried out to figure out whether there was any predominant patterns among the graphs. General topics and patterns were discovered that were consolidated into the possible outcomes of this exploratory study.

While using NASA in these higher education networks, we highlighted the catchphrases used in the network and numbered them based on their usage frequency during discussions held among the members. When the tool gathered a similar pattern across all four networks, it made no distinction between gathering catchphrase words used in the discussions and the number of times that specific catchphrase word was used, the approach had to be friendly and non-judgmental. By deciphering the results gathered from comparisons of the four networks, we can understand the patterns hidden in their interactions.

8 Results

The few noteworthy patterns that were found from the SNA were the impact of college culture, the utilization of online social networking, the formation and enhancement of various online communities within networks, exponential growth in building online relationships, associations or correspondence with others, interactions with students or HEPs, an interdisciplinary framework model of interactions, and implications for the lack of exploration, and privacy issues. Every pattern had an arrangement of the particular trend that had been recognized through this exploration.

8.1 Impact of Culture on the Use of Social Networking

As indicated by our social network graphs, we analyzed the age and demographic parameters and finally, we observed that because of the larger demographics and maximum range of age spectrum, almost all the students were admitted to the Web. Thus, it is mandatory for all the HEIs to utilize online social networking. Social graph patterns also depict the specific demographics at their respective HEI that were questionable, unexposed, and extremely uncertain when it came to the use of social networking in academics. We have seen these social networks as the *second home* for tutoring and long range interpersonal communication on Facebook (see footnote 1), which is useful for understudy achievement. The research found that three out of the four networks saw that online social networking in higher education was crucial to holding students inside and outside the HEI.

The increasing demand and fame of these social platforms give us several options for communicating with students, HEPs and educators. Social interactions can become key contributors to successful teaching and learning. We believe that a better quality of learning and growth of students can only be achieved through interaction. Whether the interaction is between students and students or students and faculty or any other, each communication or interaction contributed to the overall quality of and potential impact on students' learning. The current results in Figs. 1 and 2 depict that each student in the group is using Facebook (see footnote 1) with a different time span. Another research question that arises from this is how to determine the interaction (either positive or negative) as perceived by the student. This can foster communication.

Although designed for social uses, SNS appears to have transitioned to other areas of teenage life, including HE. Our study (Figs. 3, 4, and 5) found that in most groups, both genders use Facebook (see footnote 1). Males are using Facebook (see footnote 1) for education more than females. It was found that three-quarters of all members between the ages of 18 and 44 were using Facebook (see footnote 1). This gives rise to the research question why are there more male members than female members. This future research topic needs further analysis with regard to how gender, education, and Facebook (see footnote 1) interact. It has become more than a social networking platform for students; it has become a resource for information. The results from this study show that further work is needed to make a greater connection on Facebook (see footnote 1) that can be implemented within academia.

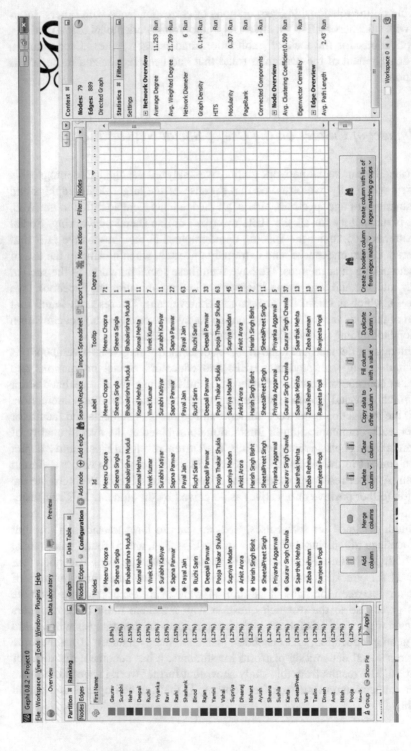

Fig. 1 These statistics show the list of members (the majority are students) and their participation percentage in the group of higher education institutions using Gephi (see footnote 16) app

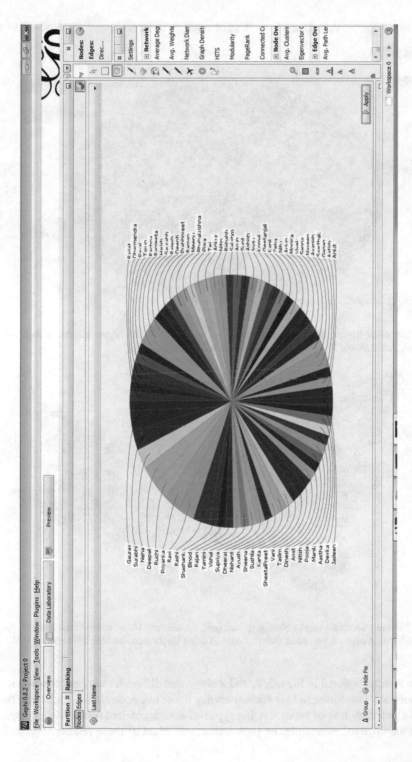

Fig. 2 The type of relationship that exists is presented in tabular form for the higher education group on Facebook (see footnote 1)

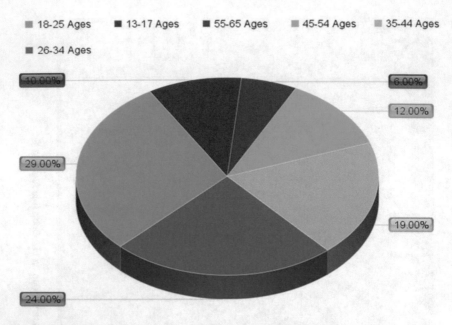

Fig. 3 Pie chart depicts the age demographics of the users of the higher education page on Facebook (see footnote 1) using Gephi (see footnote 16) software tool

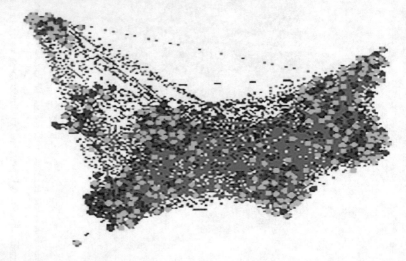

Fig. 4 Network depicting *gender demographics* (*red*, female) and (*blue*, male) of the users of the higher education page on Facebook (see footnote 1) using Gephi (see footnote 16) software tool

The results depicted in Figs. 6, 7, and 8 show that different members or students use Facebook (see footnote 1) in their everyday lives. The nodes in the social graphs are connected by one or more specific types of interdependent relationships, such as friendship, ego, commenter or post author, etc. Most of these relationships fall

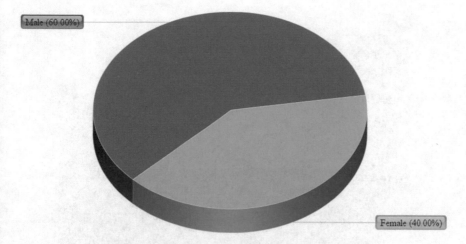

Fig. 5 Chart depicts *gender demographics* (*red*, female) and (*blue*, male) of the users of the higher education page on Facebook (see footnote 1) using Gephi (see footnote 16) software tool

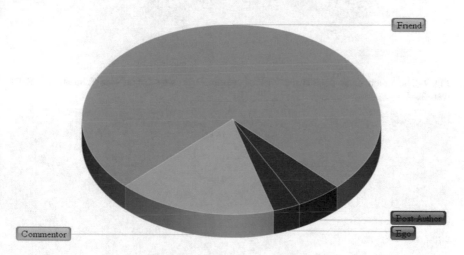

Fig. 6 The pie chart depicts the type of *relationship* using Gephi (see footnote 16) software tool

under friendship type and interaction types (comment or posts or likes). This shows that Facebook (see footnote 1) is a popular communication tool for college students to discourse with their classmates and promote even greater academic achievement in populations who dislike face-to-face interactions. Facebook (see footnote 1) is mass communication, and colleges can use Facebook (see footnote 1) to augment a student's educational experience.

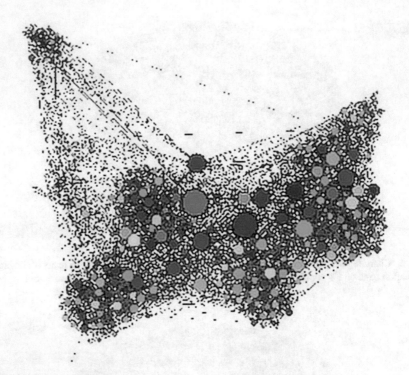

Fig. 7 The graph chart depicts the type of *relationship* using Gephi (see footnote 16) Software Tool

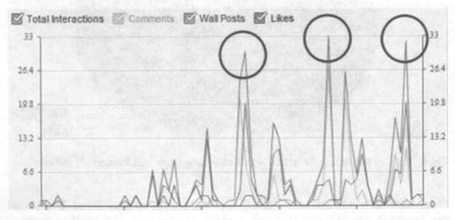

Fig. 8 The spike depicts the *interactions type* (likes, comments, posts, likes of the users of the Facebook (see footnote 1) page using the WolframAlpha app)

The student's initiative shall be a key that is not only desirable, but imperative for an "interactive environment," because these environments are effective when students write and post. These SNS are public, and thus available for study and pedagogical use. The result of our studies in Fig. 9 shows the level of interactions and Fig. 10 shows the most active members (red dots), less interested members (blue dots),

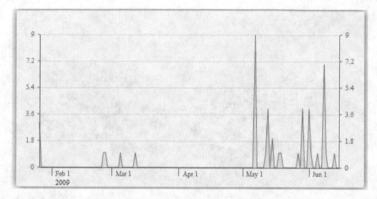

Fig. 9 The chart statistics depict the level of *interactions or discussions* by the users. (Source WolframAlpha app)

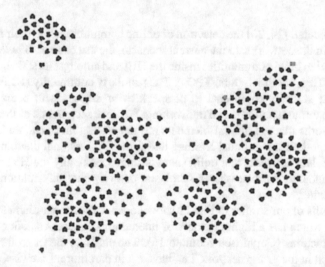

Fig. 10 Graph depicts the *active users* in the group with *red* (most active), *blue* (less active), and *green* (non-active) using Gephi (see footnote 16) software

and non-interactive (green dots). Their interaction level is high in the group either through comments, posts, messages or likes. A large percentage of the members interact most of the time because it is important to be actively involved with their classmates on Facebook (see footnote 1).

8.2 Community Enhancement

We have seen that by using online social networking, which nowadays has become an important tool for the HEIs, there is a consensus that social online networking could possibly produce more and more associations or bonding. We studied and

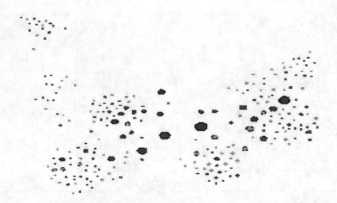

Fig. 11 Comparing the *centrality* of students in the group, using Gephi (see footnote 16) software

explored in detail [14, 23] the detection of online communities and their mesoscopic properties in the networks. In the current scenario, the use of *social media* in HE has the potential to build communities inside the HEI and amongst its stakeholders (e.g., students, HEPs, administrators, TPOs). The scholarly community (HEPs) could be enhanced if students could chat to or speak to or discuss with peers and HEPs (or faculty) through online social networking. We have seen in one of the four social graph networks that the alumni members are interfacing first-year students, which will improve their institutional learning and future professional planning. We have seen in our data networks that online networking, specifically the HEI community, connects students, but also shares scholastic or individual information and assets with each other.

The results of our studies in Fig. 11 show nodes with a larger diameter placed in the center, which has a higher number of interactions, whereas smaller nodes have fewer interactions (Gephi, see footnote 16). The students closer to the center are more central in the group network, i.e., have taken part more. Likewise, if a node is farther from the center, the student was less active and hence placed on the outer periphery of the circle. Facebook (see footnote 1) is a popular community used for both HEPs and students to keep a strong and significant basis of communication. It is one of the many sites used by academia to connect with their students. The result of our studies in Fig. 12 shows the two panels where the left panel depicts the graph metrics (the number of nodes, diameter, total edges, density, and etc.), the right panel shows the gray circles outlining the community's boundaries, and the red nodes represent the community representatives. Each community representative is accompanied by the top connected members of the community and each selected member would outline the connected members who have taken part in a common interaction in relation to a topic. This is regarded as the weight of the edges and is defined as the weight of communication between incident nodes.

The results of our studies in Fig. 13 have the left panel, which shows the terms used in the interactions. The selection of each student would outline the terms used

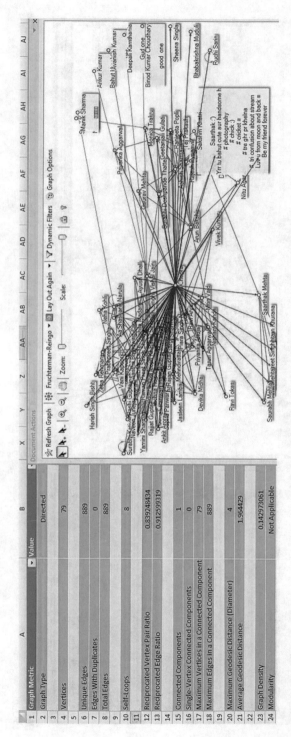

Fig. 12 Network metrics and community detection using NodeXL (see footnote 15)

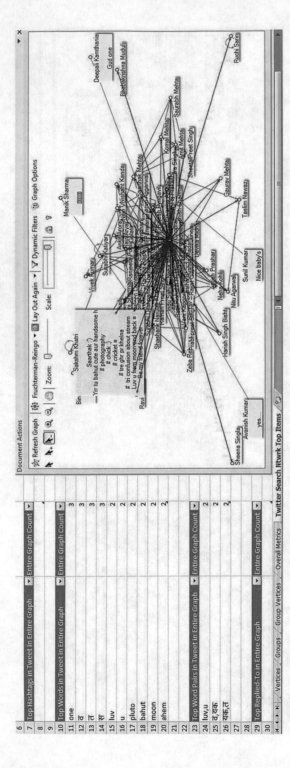

Fig. 13 Visualized *Facebook higher education* with their interactions in the *left panel* and top Tweet words in the *right panel*, using NodeXL (see footnote 15) tool

by that student. The right panel depicts the social network interaction graph of students in the group, (network graph January to June 2015), using NodeXL (see footnote 15).

8.3   Continuous Associations with Students

For the question, "How could SM help in students' academic progress?", the patterns in four social graphs showed that the students were discussing the college and other academic-related events, thus keeping them well-informed and engaged with the HEI, can be ascertained by using technology. They can easily and successfully increase the productivity of their stakeholders and reach out to their students efficiently and effectively. The Web 2.0 tools such as Facebook (see footnote 1), Twitter (see footnote 9), YouTube (see footnote 5), etc., have given HEIs a successful stage for student outreach in a way that was lacking previously. By using SM in higher education, most of the HEIs can build a more grounded association between students, HEPs, and other stakeholders. The purposes of *social media*, specifically in HEI, include connecting students, framing lines of communication with other streams of students, and sharing information that will make interaction more successful. So far, we have analyzed that the Facebook (see footnote 1) online social networking site has been observed to be a more valuable and specialized instrument in sharing information and dates with our students. It is a second home where the students can complain or express their feelings where they are not investigated or judged.

In spite the HEI perceiving that online social networking was a valuable web tool to impart, they additionally inferred that it has to be partnered with physical classroom associations. We concurred that Facebook (see footnote 1) could be utilized as a specialized instrument with the objective of having students come in and see a faculty or any peer mentor to talk about any issues such as evaluations, non-attendance or course work that students might battle with. By and large, online social networking is a specialized web tool that is the most widely recognized and adopted by all stakeholders of the HEIs. The results of our studies in Figs. 14, 15, and 16 show the various activities (such as uploaded photos, posted links, statuses, uploaded videos, etc.) were performed by the members. The time duration and amount of each activity are shown. (Statistics were taken between September 2014 and May 2015 using WolframAlpha.)

8.4   Coordinated Effort Among Members

Another pattern that surfaced amongst the four networks was the coordinated effort amongst all the members so that they can share common efforts and communicate their events and activities with peer members. The four networks of HE on Facebook

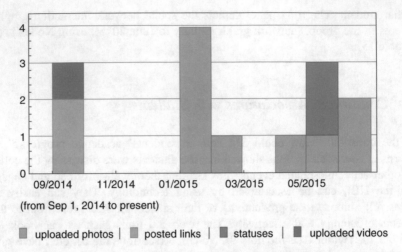

Fig. 14 These statistics show the type of events performed by network users in the form of bars

type	count	ratio	
uploaded photos	9		
posted links	2		
statuses	2		
uploaded videos	1		

64.3%

7.1%

14.3% 14.3%

Fig. 15 The statistics show the type of events performed by members in the form of pie-charts and activity performed within the group, using WolframAlpha

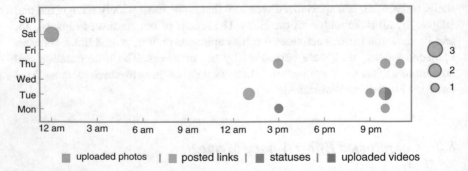

Fig. 16 The statistics show the type of events performed by members in the group with time duration, using WolframAlpha

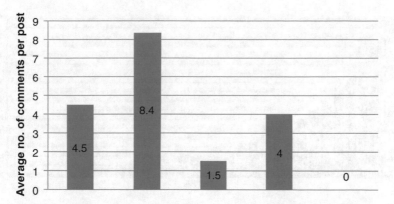

Fig. 17 To measure the average number of messages and comments from the users in the form of bar charts, using WolframAlpha (see footnote 13) and Gephi (see footnote 16)

Fig. 18 To measure the average number of messages and comments from the users based on color intensity (*red color* are the most active ones, using WolframAlpha (see footnote 13) and Gephi (see footnote 16)

(see footnote 1) gave an account of the quantity of preferences or hits they obtained on their Facebook (see footnote 1) page. These online generated reports permitted staff to team up at college social events, athletics occasions, or student activities, thus keeping students connected with their institution and helping them to team up with different classmates through online social networking and brought more communication within students, which turned out to be so overwhelming, but rather more sensible for HEI professionals.

The results of our studies depicted in Figs. 17 and 18 show the average number of comments or messages by the students in the HEI group, using the LikeAnalyzer app (see footnote 17). Gephi (see footnote 16) predicts the nodes (with different colors and sizes) showing the number of messages sent by the users and the size of

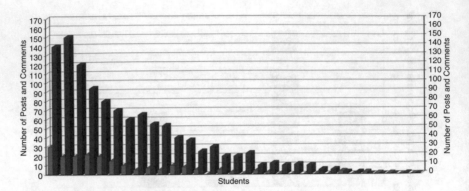

Fig. 19 The activity distribution chart in the form of bar charts, using WolframAlpha (see footnote 13) and Gephi (see footnote 16)

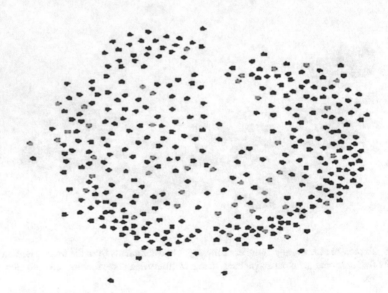

Fig. 20 The activity distribution chart in the *network form (blue, comments; green, Posts; red, inactive)*, using WolframAlpha

the node corresponding to their centrality/leadership in the discussions. The results of our studies in Figs. 19 and 20 depict the activity distribution by students' posts or comments in the HEI group, using the LikeAnalyzer app (see footnote 17), and Gephi (see footnote 16) produced similar patterns in blue (comments) and green (posts).

9   Challenges and Implications

Through research analysis, it was apparent that there was a feeling of hesitation with online social networking from most of the members of HEIs because of the lack of research on using SM in higher education. We recognize that it is too early to implement the results of SNAs patterns practically in HEIs with respect to students academic achievements. A few researchers concluded that long-range interpersonal communication on Facebook (see footnote 1) with the student was not useful to their career achievement. The effect that social media could have on student academic grades is still exceptionally indistinct to a hefty portion of the heads of the HEIs. Many HEI heads, for example, have fears about the utilization of online social networking in view of the lack of security, privacy, or secrecy that SNSs generally offer. We understand the reluctance of HEIs with regard to Facebook (see footnote 1) or other online social networking sites, as we do share many pages and there may be personal or confidential information that we do not want other students or members to know, or another individual to see, so restrictions are a primary concern. Instead of utilizing public platforms, for example, Facebook (see footnote 1) or Twitter (see footnote 9), HEIs should focus on their own local or internal social networking environments that can be used in academic and other events such as projects to connect their students or stakeholders online. These substitute internal systems are being considered and used by the HEIs' stakeholders. The built-in authentication system can protect stakeholders' rights and privacy that most HEIs are worried about. Even with the widespread use of SM in higher education, we still have some questions, such as, does an internal or external social networking platform have the security parameters that we require? Do HEIs have the technical manpower to bolster? How much control would HEIs be able to have over social networking sites so that we can keep the students and college safe? In [24], we suggested a few recommendations for the things that are still preventing us from pushing ahead on social networking platform confidentiality issues, well-being issues, and security issues. This study was restricted to four higher education-related social networks on Facebook (see footnote 1) whose users have been analyzed on the basis of a few research questions that were based on parameters such as age, demographics, total time spent, engagement factor, etc., that precisely explained the role of social media in academia. One constraint of our study is that the size of the all four networks differed.

10   Scholarly Significance

This research conducted on the four Facebook (see footnote 1) social networks of the HEIs provided a user's perspective on online social networking and its significance in these networks. This study offered points of view in the regions of collaboration (sharing), community formation [14, 23], culture, coordinated effort, communication, and challenges that accompany new technological advancements and innovations. One of the viewpoints obtained by looking at these specific HEI networks is that all the networks display the particular culture of an HEI and its stakeholders (students, faculty, etc.) have the potential to adopt technology or online social networking into their academic and social communication plans. All the programs at HEIs require their stakeholders to have technical gadgets, for example, PCs, PDAs, or web access, in their life 24×7, which might be impeded. Numerous HEI students and other stakeholders depend on this technologies-driven correspondence such as emails within and outside the campus. HEPs must consider these technological constraints when building up their key strategic academic and learning plans.

Another significant factor we found in the study was the strong role of online social networking as a communication tool. We also concluded that the maximum number of users in all four social networks of Facebook (see footnote 1) took part in at least one of the ongoing discussions and repeatedly conceded that online social networking is a great platform and tool for rapidly involving and communicating with students of higher education. These online social platforms permit students to assess their semester projects, courses, activities, or events at an HEI that takes into account prompt feedback. Continuous interaction through online social networking is the means by which students are drawn in to each other, HEPs, administrators, other staff, members, and finally the institution itself. The HEIs must embrace and adopt this reality. A vital implication that surfaced and must be recognized is that most of the HEIs are reluctant to adopt new technology in academia with a little research. In most of the cases, the online social networking is seen as a powerful social tool that may not be adequate for the academic community to embrace new technological techniques. HEIs need information and exploration to support new technology or reject it. We found that in each network there were a few users who were isolated and a sense of hesitation of users to express their thoughts. There are also other challenges such as privacy and exposure issues, and the lack of confidentiality while connected with a number of public online social networking outlets. A few questions related to online social networking in higher education may be answered in our future study. Those unanswered questions are: Does online social networking have the security parameters that HEIs need? Can HEIs bolster online social networking with the existing academic curriculum or within the current technological framework? Last, how much control can HEIs have to keep students, other stakeholders, and its institutional data safe?

11 Direction for Future Study

This study is focused on the route toward further study in drawing conclusions on and making comparisons among students, HEPs administrators, staff, and also management perceptions of online social networking in HE. Big data mining could be done by using predictive and qualitative tools to make subjective strategies and to understand students and other HEI-related workforce engagement and perceptions with the specific end goal of making observations on online social networking and its impact on the academic environment. The possible outcomes could then be compared with the present study of the perceptions of HEI stakeholders.

12 Conclusion

This study was conducted with the specific goal to obtaining the best perspective of the users, who were the members of higher education social media networks and the effect they had on the academic environment. Using NASA on the four short-listed Facebook (see footnote 1) social media networks, a few common observations were made that included the impact of the utilization of social media on academia, the exponential increase in smaller collaborations among different HEI stakeholders, virtual correspondence with the students, and the challenges of confidentiality and privacy issues. The viability and restrictions of online social networking in higher education have been exposed by this research. It proposed a typical thought that social media and the academic community can go hand in hand, as studied and examined for the four social networks. The essentials of research into online social networking in higher education may affect the future of the education industry with regard to communication, collaboration, and academic accomplishments.

References

1. Mamta, M., Meenu, C.: Social media networks (SMN) an eye: to envision and extract information. Int. J. Adv. Res. Comput. Sci. Softw. Eng. **4**(2), 1037–1041 (2014). http://www.ijarcsse.com/docs/papers/Volume_4/2_February2014/V412-0462.pdf
2. Deil-Amen, R.: Socio-academic integrative moments: rethinking academic and social integration among two-year college students in career-related programs. J. High. Educ. **82**(1), 54–91 (2011)
3. Deil-Amen, R.: The "Traditional" College Student: A Smaller and Smaller Minority and Its Implications for Diversity and Access Institutions (2011)
4. Deil-Amen, R.J., Rios-Aguilar, C.: Beyond getting in and fitting in: an examination of social networks and professionally-relevant social capital among Latina/o University students (2012)
5. Heiberger, G., Harper, R.: Have you Facebooked Astin lately? Using technology to increase student involvement (2008)
6. Higher Education Research Institute: College freshmen and online social networking sites (2007). Retrieved 25 October 2011
7. Hughes, K.L., Karp, M.M.: O'Gara, L.: Student success courses in the community college: an exploratory study of student perspectives. Community Coll. Rev. **36**(3), 195+ (2009)
8. Junco, R.: The relationship between frequency of Facebook use, participation in Facebook activities, and student engagement. Comput. Educ. **58**, 162–71 (2011)
9. Junco, R.: First semester students and sophomores spend more time on Facebook. Social Media in Higher Education (2011). Retrieved from 3 July 2011
10. Junco, R.: Too much face and not enough books: the relationship between multiple indices of Facebook use and academic performance. Comput. Hum. Behav. (2012). doi:10.1016/j.chb.2011.08.02

11. Junco, R., Heibergert, G., Loken, E.: The effect of Twitter on college student engagement and grades. J. Comput. Assist. Learn. **27**, 119–132 (2010)
12. Karp, M.M., Hughes, K.L.: Information networks and integration: institutional influences on experiences and persistence of beginning students. New Dir. Community Coll. **144**, 73–82 (2009)
13. Karp, M.M., Hughes, K.L., O'Gara, L.: An exploration of Tinto's integration framework for community college students. J. Coll. Stud. Retent. Res. Theory Pract. **12**(1), 69–86 (2010)
14. Mamta, M., Meenu, D., Meenu, C.: The mesocopic structural analysis of communities within facebook higher education online groups. Int. J. Inf. Commun. Comput. Technol. Jagan Inst. Manag. Stud. **III**(2) (2016). Source is available at: http://www.jimsindia.org/8i-v3-i2-community_detection_Facebook_AnalysisCRC.aspx
15. Mamta, M., Meenu, C.: To investigate relationships through text, link and spacial-temporal information in social media networks. Int. J. Sci. Technol. Res. **4**(3) (2015). ISSN 2277-8616. http://www.ijstr.org/final-print/mar2015/A-Review-Paper-On-Exploring-Text-Link-And-Spacial-temporal-Information-In-Social-Media-Networks.pdf
16. Mamta, M., Meenu, C.: Data mining: a mode to reform today's higher learning institutions through performance indicators. Cyber Times Int. J. Technol. Manag. **6**(1), 292 (2013). ISSN 2278-751. http://journal.cybertimes.in/?q=Vol6_A_T_40
17. Mamta, M., Meenu, C., The education gets the facelift by going social. Int. J. Appl. Innov. Eng. Manag. **2**(12), 50–53 (2013). ISSN 2319-4847. http://www.ijaiem.org/volume2issue12/IJAIEM-2013-12-09-017.pdf
18. Mamta, M., Meenu, C.: Social network wrappers (SNWs): an approach used for exploring and mining social media platforms. Int. J. Comput. Appl. **97**(17), 31–34 (2014). http://research.ijcaonline.org/volume97/number17/pxc3897687.pdf
19. Meenu, C., Mamta M.: Flipped classroom: the right solution to competent higher education institutions (HEIs). Int. J. Artif. Intell. Knowl. Disc. (IJAIKD) **5**(3), 1–10 (2015). http://www.rgjournals.com/index.php/ijai/article/view/690
20. Mamta, M., Meenu, C.: Network analysis by using various models of the online social media networks. Int. J. Adv. Res. Sci. **6**(1), 111–116 (2015). (0976 - 5697), Impact Factor (4.862)
21. Meenu, C., Mamta, M.: Social Media Analysis (SNA) using online students opinions on academic determination in higher education. Online Int. Interdisciplinary Res. J. (OIIRI) **5**(Special Issue), 142–154 (2015). http://www.oiirj.org/oiirj/may2015-special-issue/16.pdf
22. Mamta, M., Meenu, C.: Using mining predict relationships on the social media network: facebook (FB). Int. J. Adv. Res. Artif. Intell. **4**(4) (2015). ISSN: 2165-4050. http://dx.doi.org/10.14569/IJARAI.2015.040409. http://thesai.org/Publications/ViewPaper?Volume=4&Issue=4&Code=ijarai&SerialNo=9#sthash.zSJHXyNz.dpuf
23. Mamta, M., Meenu, D., Meenu, C.: Analysing online groups or the communities in social media networks by algorithmic approach. In: 50th Golden Jubliee Annual Convention, Organized by Computer Society of India (CSI). Springer (2015)
24. Mamta, M., Meenu, D., Meenu, C.: Social network analysis (SNA): in facebook higher education groups through nasa (network analysis software applications). Int. J. Artif. Intell. Knowl. Disc. **5**(4) (2015). http://www.rgjournals.com/index.php/ijai/article/view/714
25. Meenu, C., Meenu, D., Mamta, M.: Social Network Analysis (SNA): in Facebook higher education groups through NASA (Network Analysis Software Applications). Int. J. Artif. Intell. Knowl. Disc. (IJAIKD) **5**, 1–11 (2015). http://www.rgjournals.com/index.php/ijai/article/view/714
26. Sturgeon, C.M., Walker, C.: Faculty on Facebook: Confirm or deny? Paper presented at the Annual Instructional Technology Conference, Lee University, Cleveland, TN, 2009
27. Tinto, V.: Classrooms as communities: exploring the educational character of student persistence. J. High. Educ. **68**(6), 599–623 (1997)
28. Tinto, V.: Linking learning and leaving: Exploring the role of the college classroom in student departure. In: Braxton, J. (ed.) Reworking the Student Departure Puzzle, pp. 81–94. Vanderbilt University Press, Nashville (2000)

Extensible Platform of Crowdsourcing on Social Networking Sites: An Analysis

Shilpi Sharma and Hitesh Kumar

Abstract The evolutionary history of the human brain shows advancement in its complexity and creativity during its evolutionary path from early primates to hominids, and finally to *Homo sapiens*. This most powerful human asset known as the brain is highly capable of solving problems. When a problem arises, humans make use of their intelligence and various methods of finding the solution. No doubt they have come up with the best solutions, but many questions have been raised on how that problem is approached and how the solution is derived. The peculiar thing is that everyone has a different mechanism of thinking and comes up with different patterns of solutions. Can this pattern be mimicked by a machine where a problem can be solved by inputs from multiple individuals? Crowdsourcing and neural networks come into play in this domain. Crowdsourcing deals with the pooling of ideas by people. The more people, the wider the perspective obtained. The data given by them are processed and the field of neural networks plays a vital role in analyzing the data. These data contain various patterns and hidden solutions to many problems.

Keywords Backpropagation • Crowd funding • Crowdsourcing • Feed forward • Platform • Social network

1 Introduction

The working model is currently changing and is acquiring its new face with the evolution of crowdsourcing. Earlier, a team of experts was recruited and paid a large to solve an issue or problem. Team members worked on a problem and the solution was found. This solution was based on their knowledge and intelligence. The probability that they missed some of the essential points is high owing to the limited number of team members. If the number of team members is increased, the quality of the solution will be better. This is the idea that crowdsourcing works on.

S. Sharma (✉) • H. Kumar
Department of Computer Science and Engineering, ASET, Amity University, Noida, Uttar Pradesh, India
e-mail: ssharma22@amity.edu; hitesh2194@gmail.com

© Springer International Publishing AG 2017   299
H. Banati et al. (eds.), *Hybrid Intelligence for Social Networks*,
DOI 10.1007/978-3-319-65139-2_13

Crowdsourcing is a platform where all the brains work together and come up with the most reliable solution. Of course, it is not guaranteed that the solution will be better or worse. The main purpose of crowdsourcing is to increase the accuracy of the results in terms of skill sets to achieve a particular goal. The term crowdsourcing was first introduced in 2006 by "Wired" magazine. An example where crowdsourcing solved the problem is the Oxford Dictionary. Many English working-class people (800) clubbed together to extract words from all the books they came across and combined those words in a single book known as an encyclopedia. This approach became popular and there have been millions of submissions over half a century [8]. The logo design is also done using crowdsourcing. Various ideas and themes regarding the logo of a specific company are taken from a group of people and the logo is designed from the pooled ideas. Hence, crowdsourcing also solves the problem of logo design (http://www.inc.com/comcast/crowdsourcing-utilizing-the-human-cloud.html, blog accessed on November 2016). The website quora.com is one of the finest examples of crowdsourcing, where people from all communities make suggestions regarding the problems or discussions being held on that platform. This platform allows people discuss a problem from various dimensions (http://outsideinnovation.blogs.com/pseybold/2011/02/quora-crowdsourcing.html).

2   Background

Considering the current research trends, hybrid intelligent systems are being implemented on social networking sites. A large quantity of data is generated every second by the users of these social networking sites. The generated data are not just figures, numbers or charts to store, but it has hidden patterns, which can be a crucial source of new discoveries. Swarm intelligence and artificial intelligence play a pivotal role in the analysis and making the most of these data. Swarm intelligence is similar to artificial intelligence, where different individuals or a same group of agents are involved. Here, the term agents can be a group of software, machines or hardware devices. Swarm intelligence was born from the real world itself, for example, from ant colonization, flocks of birds, etc. Similarly, social networking sites act as colonies of different individuals (users) where different types of data can be extracted and a pattern can be established along with some concrete results [6]. As the social network circle is growing day by day, there is a need for a statistical tool and an intelligence system that will aid in extracting some of the hidden conclusions that may help mankind in some form or another. Reference [10] describes four measuring systems for the growth of social networking sites. SocialRank and Klout are two of the four mentioned systems, which precisely count a person's followers and what the crowd is diverting toward. These measuring systems help to create decision-making tools using artificial intelligence [2]. When the analysis becomes complicated, artificial neural networks (ANNs) come into play. Its three major jobs are classification, estimation, and optimization (CEO). ANN has been used in cases in which there is no mathematical formula [11].

3 Methodology

The study of crowdsourcing is done by studying various sources over the internet. Many successful cases have been taken into account to represent facts and results. Also, a MATLAB implementation is carried out in which data classification is performed on a dataset using neural networks.

4 Discussion

4.1 Crowdsourcing

4.1.1 Definition

Crowdsourcing is a technique or a method for outsourcing (obtaining) information, ideas or any other suggestions from a crowd of people. The main motto is to increase the coverage area of idea generation for a given problem. This method is extremely helpful in those fields where something (idea, suggestions, etc.) about anything (say, a product) is required from all perspectives (http://dailycrowdsource.com/training/crowdsourcing/what-is-crowdsourcing).

4.1.2 Scope

In this section, we try to predict the outcome of a future event using previous data. Work has been done in the field of crowdsourcing in the following areas:

1. Character recognition: Character recognition is in high demand under various technologies. It has become an important and integral part of various aspects. Whether it is industrial work or particular scientific work, it is required in every field to ease the work and do it in a smart way.
2. Image compression: Neural networks make use of various images when the sending and receiving of images are required. Compression comes into play so that transmission of imaging becomes easy and fast. Thus, the space required for storage is relatively small, and it is fairly easy to send or do the computation.
3. Stock market prediction: Neural networks come in handy for business analysts of the stock market. Using various parameters, the behavior of the stock markets can be predicted. Neural networks are the best tools to use to make predictions. They can estimate a large amount of information and can be very useful.
4. Traveling Salesman Problem: Neural networks also work for the traveling salesman problem (TSP). Calculating the minimum route with a predefined network is useful. Neural networks are extremely helpful in these types of applications (http://dailycrowdsource.com/content/crowdsourcing).

5. Medicine, electronic nose, security, and loan applications: Using realistic factors and applying this application to these fields, it can be utilized in various aspects of daily life. By means of previous data, a decision about the loan distribution to a particular person can be made.
6. Miscellaneous applications: As the research goes on, various other applications are being screened out and this field is turning out to be very helpful.

4.2 Crowdsourcing Principles

4.2.1 People

Finding the right crowd is an essential task before implementing any strategy in crowdsourcing. Have various aspects been taken into account, such as whether the community that is selected up-to-date with regard to the problem and ready to take part? This can be done by filtering the creators, critics, collectors, etc. (using various labels) to select a particular platform (http://www.teachthought.com/the-future-of-learning/trends-shifts/crowdsourced-curriculum-solutions-drive-student-outcomes).

Figure 1 shows an analysis of a survey that tells us about the crowd and their responses when asked about a particular product.

4.2.2 Objectives

One must be crystal clear with the objective. The final objective must be delivered to the crowd so that the result obtained is desirable and not vague. The community must also try to analyze the results in the best possible way.

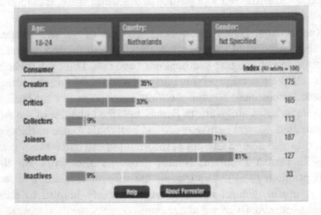

Fig. 1 Types of people

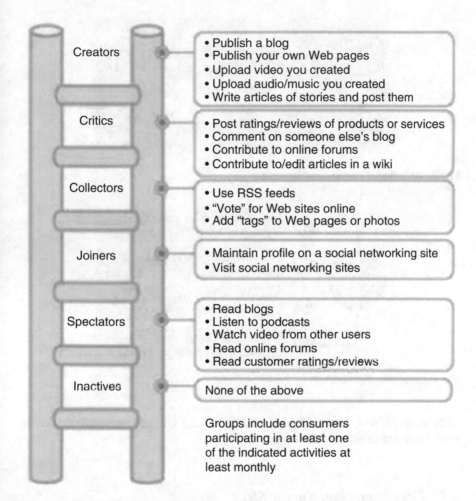

Fig. 2 Strategy ladder for crowdsourcing

4.2.3 Strategy

This includes techniques to connect to the crowd, and after that, how to obtain the maximum participation of the crowd so that the best ideas and information can be accumulated. Which peer-groups and communities should be targeted? These types of questions must be analyzed (Fig. 2).

4.2.4 Technology

Deciding on the technology used to store the responses and data is the most important step. Various applications are used to store the responses of the crowd,

Fig. 3 Technologies used in crowdsourcing

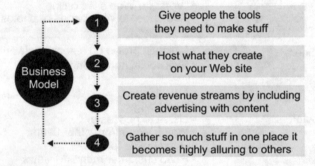

Fig. 4 Crowdsourcing model

such as WordPress, Google Forms, and SharePoint (https://openmedia.org/en/ca/crowdsourcing-principles) (Fig. 3).

4.3 Crowdsourcing: Simple Steps

In his article "Why the Power of the Crowd is Driving the Future of Business," J. Howe mentions the talented and aspiring people who are the best sources of ideas. He also advises providing them with all the required tools so that they can give their ideas and the collector of those ideas must wait for the outcomes (https://bizbriefings.com/j-howe). The content received from the crowd must always be hosted. There will be people who find the loopholes in the work. These critics play a crucial role in building the project with the least number of flaws. Gather the information and store it in one place. Make it available to everyone so that there will be multiple inferences and hence multiple conclusions. The conclusion that is best and unique can be selected. The other option is to combine these different conclusions to yield the best one (Figs. 4 and 5).

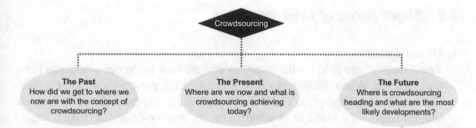

Fig. 5 The status quo of crowdsourcing

Crowd sourcing	Crowd funding
1. Sourcing: any information, service or idea	1. Sourcing: Money
2. Access: a creative manpower or workforce	2. Access: Capital for the backers
3. Helpful for business projects to include all the type of ideas from a large crowd	3. Helpful for backers to come into limelight who are talented and have the potential to provide services to the society
4. Example: quora.com	4. Example: Kickstarter or Indiegogo

Fig. 6 Differences between crowdsourcing and crowd funding

4.4 Crowdsourcing and Crowd Funding: Are These Terms Same?

Usually, crowd funding is confused with the term crowdsourcing. Crowd funding is a type of crowdsourcing that includes outsourcing money from a group of people. Here, the service providers, also known as backers, are provided with a reward or money for the services they provide. On the other hand, crowdsourcing is not exclusively associated with money. It includes various other services, either at a cost or free of charge (https://www.onespace.com/blog/2015/07/crowdsourcing-vs-crowdfunding). The major difference can be summarized from the table in Fig. 6.

4.5 Benefits of Crowdsourcing

Crowdsourcing provides a platform for turning a simple, wise internet user into a potential designer. The organization can explore talented individuals within their organization as problem solvers. The end payment is made based on the end results; hence, the customer and merchant can both be satisfied. It is a tool that converts a small bunch of customers into a flourishing business and a market (http://socialnetworking.lovetoknow.com).

4.6   Major Forms of Crowdsourcing

1. Crowd contests:
   These contests are held for the crowd to find the best worker or workforce for a specified job. When the mission is to find a specific person for a job or a task, the crowd comes into play. The task demands creativity from the crowd, which includes creating short films or advertisements, graphic design or making logos, testing software, etc. [9].
2. Macrotasks:
   Here, the crowd is being used for specific or general-purpose business work. The jobs include web page design, application development or technical content writing. The person is selected for his/her services and is paid on completion of the task.
3. Microtasks:
   As the name suggests, the most complicated tasks are broken into various subtasks that are then allocated to each person. Each microtask or unit is done and completed by the allocated person. These jobs include transaction services, medical record-keeping, and maintaining business information.
4. Crowd funding:
   Crowd funding is an essential type of crowdsourcing using social networks. The request for fund-raising is made on various social networking sites by making attractive videos or visual messages. In return, the crowd who donate money are rewarded at the end of the project.
5. Self-organized crowds:
   Here, the crowd itself is divided into various teams to complete an assigned task or challenge. The teams individually compete among themselves to find out the solution (Fig. 7).

4.7   Crowdsourcing and Social Networks

Social networks are the oxygen for crowdsourcing; more importantly, they are its backbone. It is the medium through which the whole concept of crowdsourcing started and why it has come so far.

There are various social networking sites that promote crowdsourcing. Many talented and deserving people get a chance to showcase their talents. On the other hand, the people who are looking for an effective and cost-friendly service get their work done. Social media and crowdsourcing run along the same lines assisting each other. Without social media, crowdsourcing is a herculean task. Social media act as a common bus-stop where opportunities come and go and the right person rides the right bus [3].

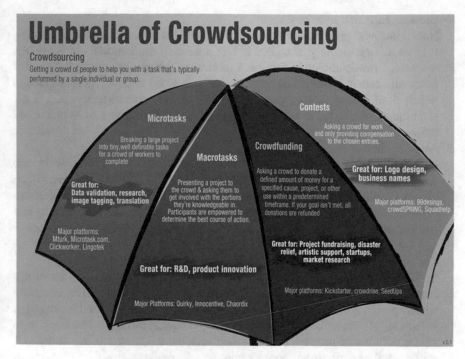

Fig. 7 Umbrella of crowdsourcing

1. Concept of connectivity and finding the right workforce for a task:
   All work needs a medium to be accomplished. Outsourcing services and ideas from a group of the correct people is an important task. Social media and crowdsourcing go hand in hand, which makes technology effective [4].
   Example: Facebook has various pages that allow investors to look for ideas posted on pages and let them contact the idea generators directly. Crowdsourcing extracts the potential service enriched with creativity and uniqueness though social media only (https://dailycrowdsource.com/content/crowdsourcing/1420-what-happens-when-crowdsourcing-and-social-media-merge).

2. Let the brain storming be done on a single spot:
   Social media act as a glue to allow individuals with the same likes and notions to be in a single spot to pool ideas with the aim of innovating. Although crowdsourcing never guarantees success, it definitely guarantees innovation (http://www.informationweek.com/10-crowdsourcing-success-stories/d/d-id/1096464).
   Twitter has also been a very favorite spot for people to outsource ideas and services. It allows like-minded people to interact and discuss a problem and come to a solution that ensures success from all perspectives [12].

3. Great ideas get the fuel:
   Great and unique ideas need a good investor and an effective plan. Social media cut the distance between the two to make an out-of-the-box idea happen.

4. Blessing for small scale business:
   Not all business organizations can afford a team to enhance their product quality.
   It takes a large quantity of capital to hire a team of professionals to get things
   working. Crowdsourcing is a tool that minimizes the cost and make things easy
   for small-scale industries. The hefty marketing cost is cut down by social media,
   which itself is a marketing guru in one way or the other. Effective and smart
   techniques are adopted to gain the customers' attention (http://www.yudkin.com/
   crowdsourcing.htm).
5. Amplification of ideas:
   It is said that great results come when all the forces work together in one
   direction. Social media with crowdsourcing is just like lemon and water, which
   blends together to produce delicious lemonade. With social media, the ideas are
   collected and a meeting point is made so that the investors and idea creators can
   work together toward a specific goal (http://www.socialmediaexaminer.com/3-
   ways-to-do-social-media-crowdsourcing).
6. The bigger the crowd, the bigger the ideas, the better the solution:
   In 2013, the well-known chips brand, Lays, used this platform to transform
   itself. Frito-Lay's share and market values were going down. The company
   came up with a plan to try new and different flavors. They started a campaign
   known as "Do Us a Flavor." They used crowdsourcing through a Facebook page
   where millions of people submitted their recipes, the winner being awarded 1
   million dollars (http://www.adweek.com/socialtimes/lays-crowdsourcing-chips/
   398334).

   Results were as follows:
   Expected response: 1 million
   Actual responses received: 3.8 million
   Number of users who installed Lay's app: 2.7 million
   Numbers of users who signed up for Facebook for this purpose: 2.2 million
   Frito-Lay's growth: 12%
   Now that is a win–win situation. This is a success story of how to use the crowd
      and increase business.

4.8   Collaborative Development

Technology giants such as Google, Facebook, and Foursquare host many events
allowing developers to work in collaboration with their official engineers to build
new and innovative apps. Recently, an idea from a blog, to build an app allowing the
user of the application to send a private message to a friend when that friend reaches
a particular location, was targeted, and after a round of 25 pizzas, 300+ coffee cups
and 15 h work, 39 crucial apps, including the above-mentioned idea, were created
(http://blogthinkbig.com/crowdsourcing-collaboration-engine-great-ideas) [5].

4.9 Drawbacks

4.9.1 Low Confidentiality

As everything is open and on social networking sites, the problem is also open to other competitive companies (http://www.businessvisions.co.uk/tag/advantages-and-disadvantages-of-crowdsourcing).

4.9.2 Success Not Always Guaranteed

It is not compulsory for there to be a win–win situation every time. The situation may get worse if proper care is not taken (http://www.cmswire.com/cms/enterprise-20/the-pros-and-cons-of-crowdsourcing-your-development-work-009327.php).

4.9.3 Multiple Hidden Costs

The work is not over after implementing the crowdsourcing. There may be multiple hidden costs that may affect the budget (https://crowdsourcedtesting.com/resources/crowdsourcing-benefits).

4.9.4 Incorrect Direction

One inexact decision related to either strategy or technology may result in a null. Hence, proper planning must be in place before pressing the accelerator (https://www.nabz-iran.com/en/content/page/lesson-one-understanding-basics-2).

4.10 Implementing Crowdsourcing Using MATLAB

One of the ways of implementing crowdsourcing is through the use of neural networks. As we know, we need a large amount of data to implement crowdsourcing. In the below implementation, we take a dataset of an Iris flower and, using its characteristics, we classify it using the backpropagation algorithm.

The dataset taken is of the flower, the iris. The data include sepal length and sepal width, petal length and petal width. Using the flower's characteristics, the type of flower is categorized by training the machine (https://visualstudiomagazine.com/articles/2013/09/01/neural-network-training-using-back-propagation.aspx). This analysis is done by training the machine and using the backpropagation algorithm.

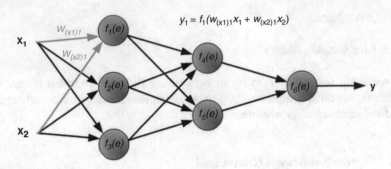

Fig. 8 The connectivity of neural network nodes

The variation of the machine learning is done by varying the number of nodes in the hidden layer.

4.10.1 Defining the Algorithm

Let us assume that a network with multiple nodes is available and that there is INPUT I with a function defined over set F. The computation is done in two phases:

1. Feed forward: The given input I is given to the network and given to the nodes for computation. The derivative is calculated and the result of the previous node is given to the forward and the end result is calculated in the final node (http:// www.fon.hum.uva.nl/praat/manual/).
2. Backpropagation: The input provided to nodes is given and the result is checked at the end. If it does not match with the required result; then the move is backward. This is done to adjust the weights and obtain the required result. This backward movement adjusts the weights or corrects the error (https://mattmazur. com/2015/03/17/a-step-by-step-backpropagation-example) (Fig. 8).
3. The algorithm is as follows:

   1. *Initialize network with random weights*

   2. *For all training cases (called examples):*

       a. Present training inputs to network and calculate output

       b. For <u>all layers</u> (starting with output layer, back to input layer):

           i. Compare network output with correct output

       (error function)

           *ii. **Adapt weights** in current layer*

	A1	▼	f_x	SEPAL LENGTH		
	A	B	C	D	E	F
1	SEPAL LENGTH	SEPAL WIDTH	PETAL LENGTH	PETAL WIDTH	FLOWER NAME	
2	5.1	3.5	1.4	0.2	Iris-setosa	
3	4.9	3	1.4	0.2	Iris-setosa	
4	4.7	3.2	1.3	0.2	Iris-setosa	
5	4.6	3.1	1.5	0.2	Iris-setosa	
6	5	3.6	1.4	0.2	Iris-setosa	
7	5.4	3.9	1.7	0.4	Iris-setosa	
8	4.6	3.4	1.4	0.3	Iris-setosa	
9	5	3.4	1.5	0.2	Iris-setosa	
10	4.4	2.9	1.4	0.2	Iris-setosa	
11	4.9	3.1	1.5	0.1	Iris-setosa	
12	5.4	3.7	1.5	0.2	Iris-setosa	
13	4.8	3.4	1.6	0.2	Iris-setosa	
14	4.8	3	1.4	0.1	Iris-setosa	
15	4.3	3	1.1	0.1	Iris-setosa	
16	5.8	4	1.2	0.2	Iris-setosa	
17	5.7	4.4	1.5	0.4	Iris-setosa	
18	5.4	3.9	1.3	0.4	Iris-setosa	
19	5.1	3.5	1.4	0.3	Iris-setosa	
20	5.7	3.8	1.7	0.3	Iris-setosa	
21	5.1	3.8	1.5	0.3	Iris-setosa	
22	5.4	3.4	1.7	0.2	Iris-setosa	
23	5.1	3.7	1.5	0.4	Iris-setosa	

Fig. 9 Excel sheet dataset

4.10.2 Steps of Implementation

1. Defining the data set:
   The data comprise sample data and test data taken from the University of California, Irvine machine learning repository. The data are stored in MATLAB variables such as INPUT, OUTPUT, and TARGET (Figs. 9 and 10).

In the above images, the label FLOWER NAME is categorized into three types by a numerical value, which is as follows:

Iris-setosa—1
Iris-versicolor—2
Iris-virginica—3

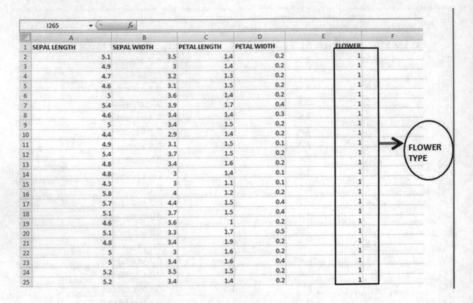

Fig. 10 Excel sheet dataset (flower names given a numerical value)

2. Creating variables in MATLAB (Fig. 11).

In the command window, the variables INPUT, OUTPUT, and TEST are being created. Values as per Excel values are copied into the MATLAB variables. These variables are then converted to double before manipulation (Fig. 12).

3. Creating neural networks using MATLAB's nntool command

Step 1: Open nntool (Fig. 13)

Step 2: Import the variables to the MATLAB database using the import button in the nntool interface (Fig. 14)

Importing the variable INPUT as input data. Similarly, import the OUTPUT variable into the target data

Step 3: Create a new neural network and import the values into its fields (Fig. 15)

Step 4: Train the network with INPUT and OUTPUT variables (Fig. 16)

Step 5: Training progress (Fig. 17)

Step 6: Using this trained machine, we can classify the data. For experimental purposes, we will use the variable TEST here.

4.10.3 Output (Mean Squared Error)

The output is achieved by the machine by inputting a dataset using the sim() function of MATLAB. The trained machine is given a set of values, the output retrieved matches the expected values, and the deviation is calculated and has been represented above by the performance graph. The output is in Figs. 18, 19, and 20.

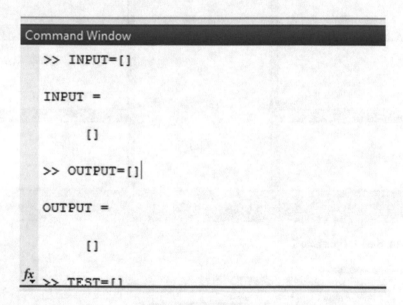

Fig. 11 Excel sheet dataset (MATLAB variable)

In the above output, the test data set were inputted with sepal (length and width), petal (length, width), and the above output categorized it according to the numbering:

Iris-setosa—1
Iris-versicolor—2
Iris-virginica—3.

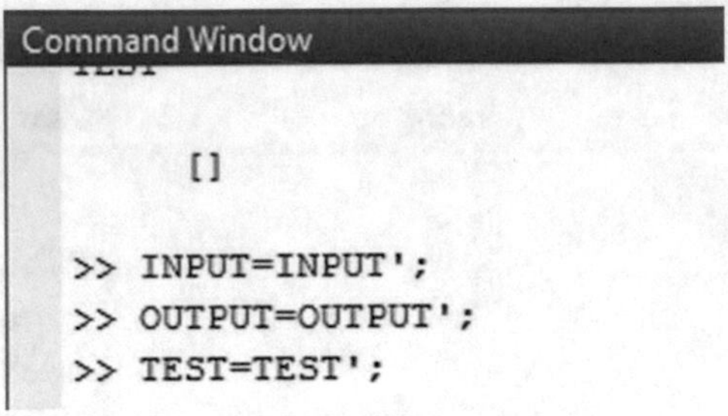

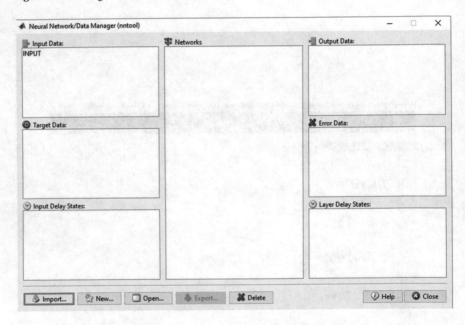

Fig. 12 Converting them to double

Fig. 13 Step 1: Open nntool

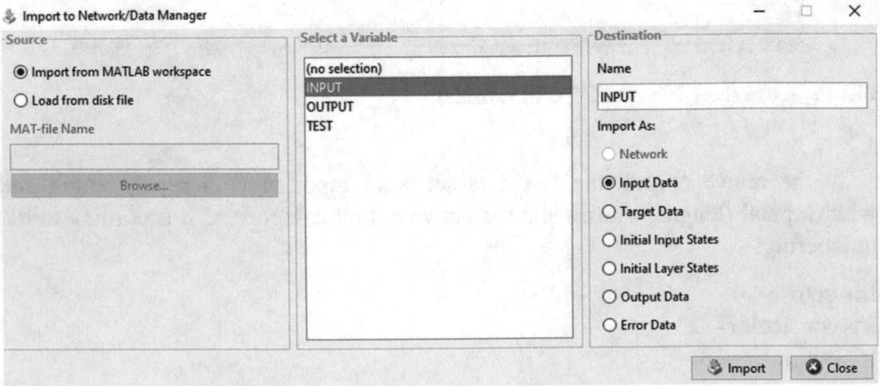

Fig. 14 Step 2: Import variables

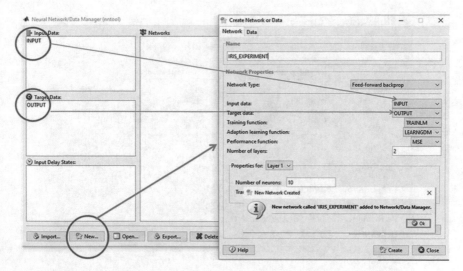

Fig. 15 Step 3: New neural network

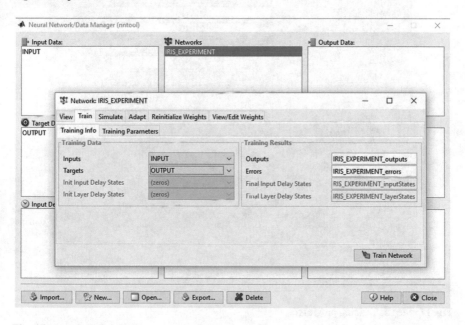

Fig. 16 Step 4: Training data

5 Results

Clearly, we can see that the trained machine, when given four input parameters of the flowers, classified it correctly using an ANN. Hence, this same concept can be used in crowdsourcing where the data can be extracted from the crowd and

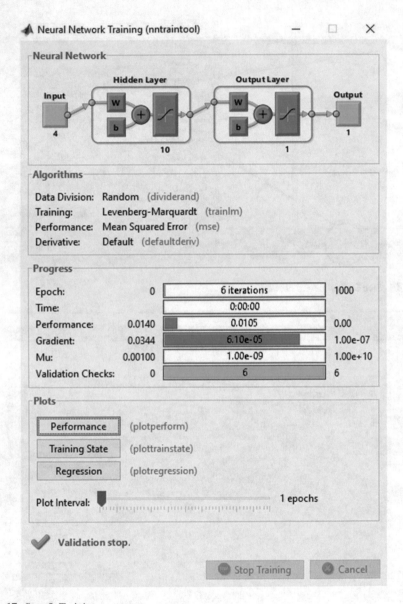

Fig. 17 Step 5: Training progress

classification can be done with particular class labels. This same concept can also be used on social networking sites by analyzing the type of users and their current preferences. Nowadays, crowdsourcing and social networking sites have a one-to-one relationship. The survival of crowdsourcing totally depends on the platform on which it is implemented. From the above discussion, it is clear that applying the

```
Command Window
>> A=sim(IRIS_EXPERIMENT,TEST)

A =

  Columns 1 through 12

    1.0068    1.0084    1.0225    1.0092    1.0045    2.1688    1.9909    1.9523    1.9612    2.1551    2.8281    2.9834

  Columns 13 through 15

    2.9858    2.9990    2.9998
```

Fig. 18 The output of the trained machine (classification)

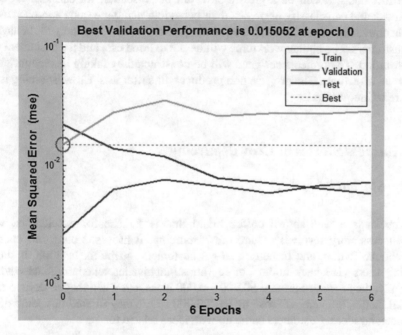

Fig. 19 The figure showing the validations and training curve of the trained machine using mean squared error

SEPAL LENGTH	SEPAL WIDTH	PETAL LENGTH	PETAL WIDTH	CALCULATED (X)	TRAINED MACHINE OUTPUT(Y)	(Y-X)^2	
5.4	3.7	1.5	0.2	1	1.006800222	4.6243E-05	
4.8	3.4	1.6	0.2	1	1.008418851	7.0877E-05	
4.8	3	1.4	0.1	1	1.022499501	0.000506228	
4.3	3	1.1	0.1	1	1.009235193	8.52888E-05	
5.8	4	1.2	0.2	1	1.004499878	2.02489E-05	
6.1	2.9	4.7	1.4	2	2.16877204	0.028484001	
5.6	2.9	3.6	1.3	2	1.990944294	8.20058E-05	
6.7	3.1	4.4	1.4	2	1.952312004	0.002274145	
5.6	3	4.5	1.5	2	1.961197516	0.001505633	
5.8	2.7	4.1	1	2	2.155101592	0.024056504	
6.5	3.2	5.1	2	3	2.828099379	0.029549823	
6.4	2.7	5.3	1.9	3	2.983357447	0.000276975	
6.8	3	5.5	2.1	3	2.985776676	0.000202303	
5.7	2.5	5	2	3	2.999038575	9.24339E-07	
5.8	2.8	5.1	2.4	3	2.999794165	4.23681E-08	
						sum=	0.087161242
						MSE=(SUM/15)=	0.005810749

Fig. 20 Mean squared error calculation

best strategy and model yields the best result, which is better than with the use of conventional methods. The project cost is also reduced and the innovation factor is high.

6   Conclusion

In the future, crowdsourcing will be in great demand. With its upcoming of projects with low costs, it will be a great tool. From our research, we can take crowdŠs feedback and can classify or predict their behaviour using the same approach as we did in above flower set. By using the same approach, we can also study the behaviour of customers. The human workforce will be put to good use, and the best talent will be awarded. The projects designed will be constructed by taking in account views from all directions. The projects or a product will suffer less. Crowdsourcing is the future of innovation.

7   Success Stories of Crowdsourcing

7.1   Starbucks

Starbucks is a well-known coffee brand that is famous for excellent service. Its success story revolves around crowdsourcing. It has used platforms such as Facebook, Twitter, and Instagram to get customers' attention. In 2008, it started taking users' feedback about coffee (https://archive.ics.uci.edu/ml/datasets/Iris). Initially, they received around 50,000–60,000 ideas, but when they took help from social media, the numbers rose up to 200,000 where they filtered out some of the spectacular ideas. Customers came up with ideas related to:

(a) Services
(b) Products
(c) Coffee shop infrastructure

What did Starbucks get in return?

1. Unique ideas and information:
    After using the crowdsourcing technique, Starbucks had a huge number of ideas to go out and implement. The reason for implementing these ideas came from their customers, who only want their well-being (https://digit.hbs.org/submission/my-starbucks-idea-crowdsourcing-for-customer-satisfaction-and-innovation).
2. Customer involvement:
    It feels great for a customer if they are given importance. The crowdsourcing makes the customer the real king, as all the ideas come from them.

They also get in contact with the organization to take part in their activities (https://socialmediarsm.wordpress.com/2015/10/20/mystarbucksidea-com-starbucks-succesful-crowdsourcing-initiative). Satisfying a customer is the ultimate aim of the organization. This can be easily done by implementing what the users want. Hence, crowdsourcing is the ultimate plan (http://www.campaignlive.co.uk/article/1317052/mcdonalds-rolls-crowdsourced-burgers).

7.2   McDonalds Burger Innovation Campaign

McDonalds started a burger builder campaign. They allowed their valuable customers to submit their choice of burger. The data were collected in which the customer submitted their burger wish. A country-wide vote was carried out and the best burger was brought into the stores (http://www.businessinsider.in/These-Are-The-5-Burgers-Brits-Said-McDonalds-Should-Make-Next-And-McDs-Is-Doing-It/articleshow/44827176.cms); (http://www.prnewswire.com/news-releases/burger-crowdsourcingat-mcdonalds—-pretzelnator-is-giving-the-classics-a-national-touch-148679665.html).

The benefits they received are as follows:

1. McDonalds did not have to do any brainstorming; they received all the unique ideas through crowdsourcing.
2. The customers were delighted to submit their choice of burgers that they wanted to see in the stores.
3. There were a million responses and McDonalds had a treasure chest of ideas, which they could use at any time.

7.3   Lego

Lego is famous toy company and one of the best examples of crowdsourcing. It allowed people to submit any idea. The idea for the product was received and the voting was done on the website only, and it had to get the maximum number of votes to be implemented. This is the finest example of a crowdsourcing success story as it encourages people's participation. This brings the company into the limelight.

The company rewarded a percentage of the revenue earned, to the person who brought in the best idea.

7.4   Greenpeace Advertisements

In 2012, there were many environmental saving advertisements and campaigns. Greenpeace used the concept of crowdsourcing to make people aware in the form of advertisements. They outsourced various quotes and ideas from people and showed them in advertisements. These quotes were ironic and funny, which illuminated a message to make people aware of the environment. Greenpeace's agenda was to stop the arctic meltdown of ice (http://digitalsparkmarketing.com/crowdsourcing-design).

7.5 Anheuser–Busch

Anheuser–Busch (AB) is one of the best-selling beer brands in the USA. It gained an edge over its competitors when it planned a collaboration with 25,000 customers. They used the crowdsourcing platforms to obtain views and ideas on new beer tastes and flavors. The ideas were related to tastings the beer. More than 30,000 photographers benefitted from this campaign as they were providing highlights of all the activities[1].

7.6 GenMil (General Mills)

GenMil is a firm for food processing that took a giant step in food processing innovation by collaborating with other ventures. Its main aim was to improve the processing steps to benefit both the producer and the consumer (https://digit.hbs.org/submission/anheuser-busch-innovating-beer-through-crowd-sourcing). In its campaign, it collected ideas and suggestions for each level of production, which includes:

(a) Improved packaging
(b) Improved distribution
(c) Improved food ingredients

References

1. Ambani, P.: Crowdsourcing new tools to start lean and succeed in entrepreneurship: entrepreneurship in the crowd economy. Crowdfunding Sustain. Entrep. Innov. 37 (2016)
2. Atlas, D., Çilingirirturk, A.M., Gulpinar, V.: Analyzing the process of the artificial neural networks by the help of the social network analysis, pp. 80–91 (2013)
3. Biella, D., Sacher, D., Weyers, B., Luther, W., Baloian, N., Schreck, T.: Crowdsourcing and knowledge co-creation in virtual museums. In: CYTED-RITOS International Workshop on Groupware, pp. 1–18. Springer International Publishing, Berlin (2015)
4. Bongard, J.C., et al.: Crowdsourcing predictors of behavioral outcomes. IEEE Trans. Syst. Man Cybern. Syst. Hum. 43(1), 176–185 (2013)
5. Doan, A., Ramakrishnan, R., Halevy, A.Y.: Crowdsourcing systems on the world-wide web. Commun. ACM 54(4), 86–96 (2011)

6. Dragoni, M.M.N., Marrafa, L.B.A., de Nicola, S.: Social networks and collective intelligence: a return to the Agora. Soc. Netw. Eng. Secure Web Data Serv. **88** (2013)
7. Gellers, J.C.: Crowdsourcing sustainable development goals from global civil society: a content analysis (2015)
8. Howe, J.: The rise of crowdsourcing. Wired Magazine. Accessed at https://www.wired.com/2006/06/crowds/ (2006). Accessed Nov 2016
9. Li, J., et al.: Learning from the crowd with neural network. In: 2015 IEEE 14th International Conference on Machine Learning and Applications (ICMLA). IEEE, New York (2015)
10. Melin, P., Castillo, O., Kacprzyk, J. (eds.) Nature-Inspired Design of Hybrid Intelligent Systems, vol. 667. Springer, Berlin (2016)
11. Peng, X., Babar, M.A., Ebert, C.: Collaborative software development platforms for crowdsourcing. IEEE Softw. **31**(2), 30–36 (2014)
12. Schimak, G., Havlik, D., Pielorz, J.: Crowdsourcing in crisis and disaster management—challenges and considerations. In: International Symposium on Environmental Software Systems. Springer International Publishing, Berlin (2015)

Index

© Springer International Publishing AG 2017
H. Banati et al. (eds.), *Hybrid Intelligence for Social Networks*,
DOI 10.1007/978-3-319-65139-2

Printed in the United States
By Bookmasters